THE BRAIN

An Introduction to Functional Neuroanatomy

THE BRAIN

An Introduction to Functional Neuroanatomy

Charles Watson
Curtin University
Perth, Australia

Neuroscience Research Australia
Sydney, Australia

Matthew Kirkcaldie
University of Tasmania
Hobart, Australia

George Paxinos
University of New South Wales
Sydney, Australia

Neuroscience Research Australia
Sydney, Australia

ELSEVIER

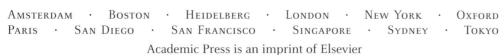
Academic Press is an imprint of Elsevier

Academic Press is an imprint of Elsevier
32 Jamestown Road, London, NW1 7BY, UK
30 Corporate Drive, Suite 400, Burlington, MA 01803, USA
525 B Street, Suite 1800, San Diego, CA 92101-4495, USA

First edition 2010

Copyright © 2010 Elsevier Inc. All rights reserved

Cover design by Lewis Tsalis
Book design by Lewis Tsalis

No part of this publication may be reproduced, stored in a retrieval system or transmitted in any form or by any means electronic, mechanical, photocopying, recording or otherwise without the prior written permission of the publisher

Permissions may be sought directly from Elsevier's Science & Technology Rights Department in Oxford, UK: phone (+44) (0) 1865 843830; fax (+44) (0) 1865 853333; email: permissions@elsevier.com. Alternatively visit the Science and Technology Books website at www.elsevierdirect.com/rights for further information

Notice
No responsibility is assumed by the publisher for any injury and/or damage to persons or property as a matter of products liability, negligence or otherwise, or from any use or operation of any methods, products, instructions or ideas contained in the material herein. Because of rapid advances in the medical sciences, in particular, independent verification of diagnoses and drug dosages should be made

British Library Cataloguing in Publication Data
A catalogue record for this book is available from the British Library

Library of Congress Cataloging-in-Publication Data
A catalog record for this book is available from the Library of Congress

ISBN: 978-0-12-373889-9

For information on all Academic Press publications
visit our website at www.elsevierdirect.com

Printed and bound in China

10 11 12 13 14   10 9 8 7 6 5 4 3 2 1

Working together to grow libraries in developing countries

www.elsevier.com | www.bookaid.org | www.sabre.org

ELSEVIER   BOOK AID International   Sabre Foundation

To our children and grandchildren–Julia and Anna, Callie and Edie, Alexi and Yvette, Juno and Luka, and Cairo and Aida.

Preface

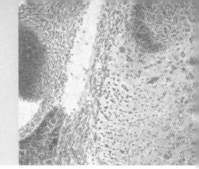

This book breaks away from two long-standing traditions of neuroanatomy teaching. First of all, it begins with a presentation of the organization of a simple mammalian brain, instead of commencing with the anatomy of the human brain. We reserve the discussion of the complexities of the human forebrain for the second half of this book. Secondly, we have introduced some new concepts of brain nomenclature based on recent data on gene expression during development, instead of the arbitrary naming system based on external features. This means that terms such as diencephalon, pons, and medulla have new and different meanings in this book. In addition, we have rejected the continuing use of confusing terms such as limbic system, reticular activating system, and extrapyramidal system. These outdated terms are popularly used in defiance of recent evidence. We do not expect these nomenclature changes to be welcomed by clinical neurologists, but we believe that young neuroscientists should be aware that it is time to embrace new concepts derived from molecular biology.

We have tried to write a book that would inspire and inform beginning students in neuroscience. We have moved to concepts revealed by modern neuroscience research. We believe this book will be useful for beginning students in biology and in the health sciences.

Charles Watson
Matthew Kirkcaldie
George Paxinos

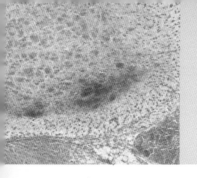

Acknowledgments

We thank our team of medical undergraduates, Megan Harrison, Aine O'Brien, and Katherine Fuller, for their outstanding work on the text and the illustrations. Dr Larry Swanson and Dr Dan Sanes gave us permission to adapt their excellent diagrams. We have also gathered ideas for illustrations from a number of other sources, which are acknowledged in the relevant legends.

We thank our many colleagues who generously donated photographic images for use in this book. Dr Hironobu Tokuno allowed us to use images from the superb marmoset brain series generated by his team at the Tokyo Metropolitan Institute for Neuroscience, and also gave us the delightful photograph of the fighting kangaroos. Dr Jan Provis provided the primate retinal images, and Dr Sandra Rees donated images of Golgi stained cortical cells. Dr Randall Moldrich gave us the striking DTI images. Dr Stacey Cole gave us the image of migrating cortical interneurons.

Professor David Ebert gave us a beautifully rendered MRI head. Stan Mitew provided a dramatic image of cellular damage in Alzheimer's disease. Michael Thompson and Dr Lisa Foa let us choose among their excellent growth cone pictures. Dr Graham Knott gave us a range of superb electron micrographs. Dr Mark George provided us with fascinating MR images.

Dana Bradford made many useful suggestions that have been incorporated into the book. Paul Watson was responsible for meticulous copy-editing.

Lewis Tsalis of Lewis Tsalis Design developed and finalized the figures, created the book cover and the book design, and carried out the book layout. As always, his work has greatly enhanced the publication.

We thank the Elsevier editorial staff, Mica Haley, Johannes Menzel, and Melissa Turner, for their patience and flexibility in seeing this project through to fruition.

Introduction

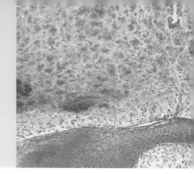

This book is a guide to the anatomy and function of the nervous system.

Almost all neuroanatomy textbooks focus on the anatomy of the human brain, probably because the authors hope that many of their readers will be medical or health science students. The problem with this approach is that the structure of the human brain is extremely complex, and a difficult point of entry into the basic structure of the mammalian brain.

We have chosen to focus initially on a simpler mammalian brain, that of the rat. We have found that this is a better way of introducing students to the major structural and functional features of the brain. All the major components of the mammalian brain are present in the rat, but the layout is much simpler. When we have covered the basic systems of the mammalian brain (Chapters 1 to 6), we introduce the reader to the wonderful intricacies of the human brain through a detailed study of the human cerebral cortex (Chapter 7).

In our approach to the presentation of the basic functions of the brain, we have focused on the fundamental role of the mammalian nervous system from an evolutionary point of view. The role of the nervous system is to ensure the survival of the individual and the species to which the individual belongs. In other words, the nervous system must ensure that the individual can obtain food and water, fend off attackers, and reproduce and care for their babies.

All of these functions depend on the ability of the nervous system to gather and analyze significant information from the external world and from the internal workings of the body. The nervous system is informed by sensory systems, which detect light, sound, touch, and other information from the external environment, or which give information about the state of the internal organs. The nervous system analyzes this information and compares it with previous experience, in order to choose what type of response is required. It is important to remember that survival in evolutionary terms is not just saving the life of the individual, but also ensuring that the species survives through the production and rearing of offspring.

The nervous system can respond to significant external and internal information in many different ways: it can cause muscles to contract; it can regulate the activity of internal organs; and it can release hormones into the blood stream. These responses might be quite simple (such as raising the arm, increasing the heart rate, or increasing secretion of stress hormones) or they may be sustained, complex behaviors with long-term goals (like building a house or child rearing). More elaborate nervous systems are generally more successful because they can generate a greater range of possibilities; the complexity of the human brain allows us to generate the widest variety of behavior of any organism.

The aim of this book is to give the reader an appreciation of the systems in the brain and spinal cord that gather and analyze information, and the systems which allow the body to respond to the challenges presented by this information.

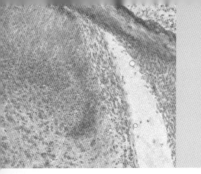

Table of contents

Chapter One Nerve cells and synapses 1
    Membrane potentials and action potentials 2
    Neurons and their connections 2
    Glia 8

Chapter Two Central nervous system basics–the brain and spinal cord 11
    Why study animal brains? 12
    The main parts of the brain 14
    External features of the brain 15
    The language of brain anatomy 18
    The spinal cord 19

Chapter Three A map of the brain 25
    Mini-atlas of the rat brain 26

Chapter Four Peripheral nerves 43
    Motor and sensory nerves 44
    Somatic and visceral motor and sensory elements 45
    Spinal nerves 46
    Spinal nerves supplying the limbs 47
    Cranial nerves 47

Chapter Five Command and control–the motor systems 55
    Command and control of skeletal muscles 56
    Areas of the motor cortex 57
    Survival skills: the hypothalamus 59
    Brainstem and spinal cord modules for control of organized movement 60
    Descending control pathways other than the corticospinal tract 61
    The role of the cerebellum in motor control 61
    The roles of the striatum and pallidum in motor control 64
    The final common pathway for all motor systems–the motor neuron 66
    Command and control of the viscera–the autonomic nervous system 68
    Command and control of the neuroendocrine system 72

Chapter Six Gathering information–the sensory systems 75
    Receptors 76
    Keeping sensory maps intact 77
    Interpretation and understanding 77

Sensory areas in the cerebral cortex 78
Vision 84
Hearing 87
Vestibular system 89
Taste 93
Smell 93
Sensory processing outside the cortex 95
An example: rolling an ankle 95

Chapter Seven The human cerebral cortex 97
The cerebral cortex–anatomy and histology 98
Guiding principles of cortical structure and function 102
The functional layout of the human cerebral cortex 103
The cerebral cortex and behavior 106

Chapter Eight Higher level functions–consciousness, memory, learning, and emotions 109
Consciousness 110
Memory 111
Sleep 117
Emotions and the amygdala 121

Chapter Nine When things go wrong–brain disease and brain injury in humans 125
Infections of the brain and spinal cord 126
Multiple sclerosis 128
Parkinson's disease 129
Stroke (cerebrovascular accident) 130
Alzheimer's disease and dementia 130
Epilepsy 132
Brain trauma and brain death 132
Mental illness 133
The tragic history of the treatment of severe mental illness 136

Chapter Ten The development of the brain and spinal cord 141
Genes and brain development 142
Early development of the brain and spinal cord 142
Regional development of the nervous system–segmentation and organizing centers 145
Formation of synapses 148
Axon guidance 149
Environmental influences on gene expression 150
Critical periods 150
Neural plasticity 151
Later processes that refine the structure of the brain 152

Continued page XII

Table of contents (continued)

Chapter Eleven **Techniques for studying the brain** 153
    Cutting thin sections of the brain 154
    Staining brain sections 154
    Cell culture 160
    Hodology: using tracers to map connections 160
    Molecular genetics 161
    Non-invasive imaging techniques 162
    Functional imaging 163
    Electrophysiology 164

Appendices 166

Supplementary reading 174

Glossary 180

Bibliography 190

Index 194

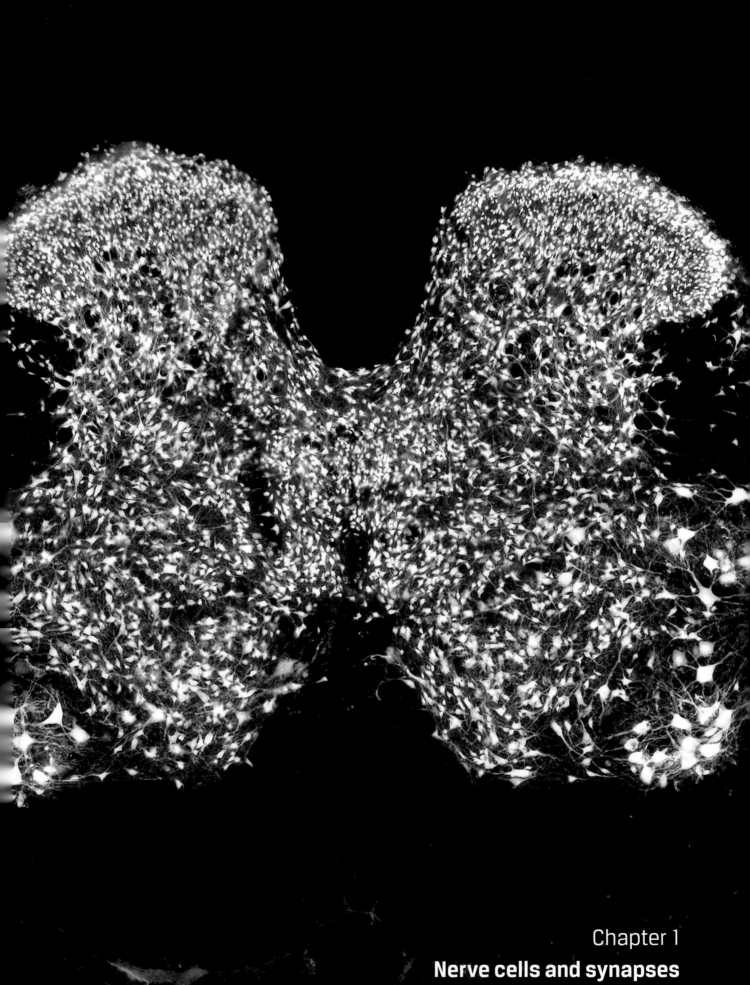

Chapter 1

Nerve cells and synapses

Nerve cells and synapses

In this chapter:

Membrane potentials and action potentials — 2

Neurons and their connections — 2

Glia — 8

The nervous system is made up of two types of cells—neurons and glia. A typical neuron has dendrites, which receive input from other neurons or from sensory receptors, and a single axon, which carries signals to other neurons or to muscles. When the dendrites are stimulated, the neuron generates an electrical signal, which is transmitted along the axon. Axons can vary in size from being very short to very long, even up to half the body length of an animal. When the electrical signal reaches the end of the axon, a chemical transmitter is released, which influences a second neuron.

Neurons come in many shapes and sizes. Some have cell bodies as large as 50 micrometers (one twentieth of a millimeter), and some are as small as 10 micrometers (one hundredth of a millimeter). For comparison, the average diameter of a human hair is 100 micrometers.

The second cell type in the nervous systems is the glial cell. Glia were originally thought to have simple nutritive and protective functions, but our understanding of their roles has expanded considerably in recent years. Glia are now viewed as functional partners to neurons, with key roles in regulating the activity of neurons and the communication between them. Various subtypes of glia protect and buffer the central nervous system against chemical changes; others provide immune surveillance or alter the properties of neuronal signaling.

Membrane potentials and action potentials

Living cells maintain their internal chemical environment by controlling the movement of ions and molecules across their membranes. Neurons can pump large quantities of ions across their membranes to produce a voltage difference (an electrical potential) between the inside and outside of the cell. This is called the membrane potential[1]. Stimulating a neuron can trigger a large, brief jump in membrane potential, called an action potential or 'spike.' An action potential at one point on the surface of a cell will set off action potentials around it in a chain reaction, spreading almost instantly across the entire cell and down its axon. Transmission of action potentials is the basis of most communication within the nervous system.

Neurons and their connections

Neurons receive stimuli on their sensitive filaments called dendrites. The membrane covering each dendrite has many tiny channels which control the flow of positive and negative ions across the membrane. Some of these ion channels are sensitive to chemical or physical stimuli, and can cause changes in the electrical charge on the membrane. If enough of these small membrane voltage changes happen at the same time, they trigger an action potential. When an action potential is triggered, this sharp, clear signal is transmitted along the axon. At its far end, the axon breaks up into a number of terminal branches. Specialized swellings on these branches are called axon terminals or terminal boutons. An action potential causes the boutons to release chemical signaling compounds called neurotransmitters. These neurotransmitters connect in a key-and-lock fashion with receptors on the target cell. Receptors are protein structures that are shaped to receive specific transmitters. The combination of the neurotransmitters with the receptors can

[1] See Appendix for more about membrane and action potentials.

Figure 1.1 Different types of neurons

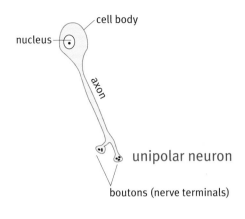
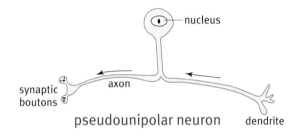
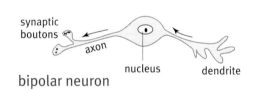
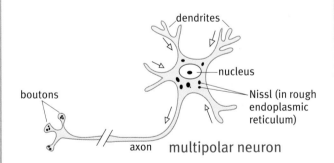

This diagram shows some of the different types of neurons found in the nervous system of different animals. Unipolar neurons have an axon but no dendrites. Pseudounipolar neurons have an axon and a single dendrite but the dendrite and axon are fused near the cell body. Bipolar neurons have an axon and a single dendrite whereas multipolar neurons have many dendrites. (Adapted from Brodal, 2004, p 7)

Figure 1.2 The parts of a neuron

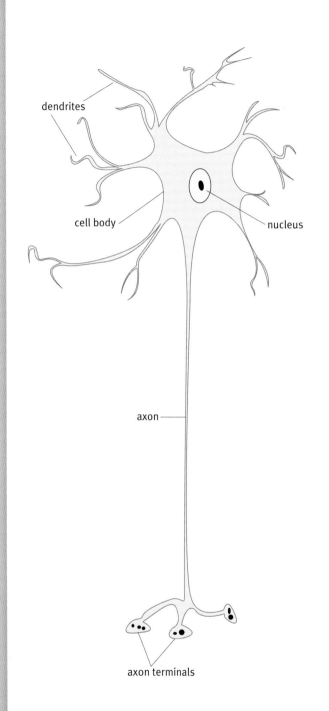

This diagram shows the basic structures that make up a neuron. The neuron cell body has many branching dendrites. A long thin axon extends away from the cell body. The axon will eventually break up into many fine branches which end in swellings called terminal boutons. The boutons make synaptic contact with other neurons or with muscle cells.

open ion channels or trigger other changes. Neurotransmitter receptors are mostly found on the dendrites of other neurons, but they are also found on the surface of muscle cells, glial cells, or gland cells. The close physical pairing between an axon terminal and a concentration of receptors on another cell is called a synapse.

Synapses

Neurons are able to receive and integrate information from multiple stimuli, and can send messages to distant regions of the nervous system. The activity of a single neuron in the central nervous system may be influenced by tens of thousands of synaptic inputs. In the cerebellum up to a quarter of a million synapses are connected to a single Purkinje cell. The axon that carries information away from the neuron may branch into hundreds or even thousands of terminals. Conversely, there are neurons dedicated to a single input, like a photoreceptor in the eye.

Synapses are just one of the influences on neuronal firing. Other influences include the activity of nearby glia, the composition of extra-cellular fluid, and the presence of circulating hormones. Direct electrical links also exist between the membranes of some neurons. To further complicate the picture, many of these types of communication can be two-way, so that an axon terminal that synapses on a dendrite might also be influenced by feedback from the dendrite. Because receptors often have effects on internal biochemistry and gene expression, the communication between

---

Synapse

A synapse is a small region of close proximity between the terminal branch of an axon and another cell. Messages are transmitted across the synapse by neurotransmitters.

How many neurons?

The rat brain contains 200 million neurons. The rhesus monkey brain contains six billion neurons. The human brain has 86 billion neurons (Herculano-Houzel, 2009).

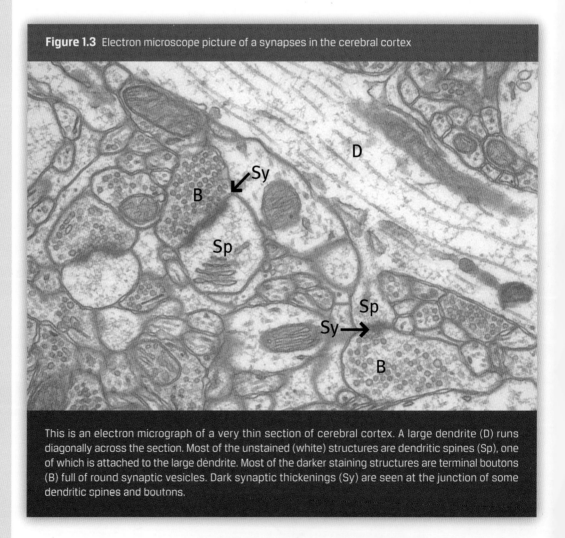

Figure 1.3 Electron microscope picture of a synapses in the cerebral cortex

This is an electron micrograph of a very thin section of cerebral cortex. A large dendrite (D) runs diagonally across the section. Most of the unstained (white) structures are dendritic spines (Sp), one of which is attached to the large dendrite. Most of the darker staining structures are terminal boutons (B) full of round synaptic vesicles. Dark synaptic thickenings (Sy) are seen at the junction of some dendritic spines and boutons.

cells in the nervous system is rich and diverse, much more so than simplistic models of neurons as integrators of their inputs. (See box, Computers and nervous systems, p10).

Synapses can be changed. Even in the adult brain, synapses may be discarded and new ones formed according to need. This process of tuning and revising connections is termed plasticity. Plasticity is the basis of developmental maturation, learning, memory, recovery from injury, and indeed every functional adaptation of the nervous system. All regions of the nervous system are plastic to some extent, but the extent of plasticity varies a great deal. Generally, the juvenile nervous system is more plastic than the adult, and the cerebral cortex is more functionally plastic than the brainstem and spinal cord.

Most communication between neurons occurs by the release of chemical neurotransmitters from axon terminals and the detection of these chemicals on the surface of cell membrane. Synapses are the spaces in which this communication takes place. They may be small and relatively open to the extracellular space, or large and enclosed, but they share common structural features. The neuron or sensory receptor that secretes a neurotransmitter is called the presynaptic cell, and the corresponding receptors are in the membrane of the postsynaptic cell. Synapses are activated by action potentials that travel to the axon terminals. The action potential causes calcium ions to enter the terminals, and this in turn activates cellular machinery to deliver vesicles full of neurotransmitter to the presynaptic membrane. The number and size of vesicles delivered, as well as the way in which they release their contents, vary greatly with cell type. Neurotransmitter molecules are released from the presynaptic memebrane and diffuse rapidly across the nanometer-sized fluid space of the synaptic cleft. The neurotransmitter binds to and activates a receptor on the postsynaptic membrane. Receptor binding is a dynamic process, with transmitters occupying active sites for only a fraction of the time they are present in the cleft. Their time for activation is limited by the action of enzymes, which break down the transmitter. In other situations the transmitters are pumped away by specialized transport systems on glia or presynaptic cell membranes. This all happens within milliseconds. For example, visual stimuli are signaled by interruptions in the steady stream of neurotransmitter flowing from the visual receptors. Retinal glia take up the neurotransmitter so fast that light flickering tens of times per second is easily noticeable.

Chemical communication between neurons and their targets can also occur outside synapses. For example, in the autonomic nervous system it is common for receptors to be activated by neurotransmitter leaking from nearby axons, or by substances circulating in the blood.

Neurotransmitters

The chemical compounds used by neurons to signal to each other are collectively called neurotransmitters. Many different neurotransmitters have been identified. Here we will summarize the action of the more common neurotransmitters.

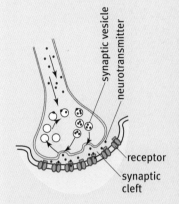

Figure 1.4
Structure of the synapse

This diagram shows the basic components of a synapse. The terminal bouton of the axon contains synaptic vesicles, each of which encloses many neurotransmitter molecules (shown here as pink dots). The neurotransmitter molecules are released into the synaptic cleft. The neurotransmitter diffuses across the synaptic cleft and binds to receptors on the post-synaptic cell membrane. (Adapted from Campbell, 1996, p 1004)

Table 1.1 A table of common neurotransmitters

Neurotransmitter	Descriptive name	Typical functions
glutamate (glu)	glutamatergic	CNS excitation
aspartate (asp)	glutamatergic	brain, spinal cord excitation
γ-aminobutyric acid (GABA)	GABAergic	CNS inhibition
glycine (gly)	glycinergic	rapid inhibition in spinal cord
acetylcholine (ACh)	cholinergic	muscle/autonomic activation; attention
dopamine (DA)	dopaminergic	reward; movement
noradrenaline (NA) [a.k.a. norepinephrine (NE)]	noradrenergic	arousal; smooth muscle control
serotonin (5-HT)	serotonergic	relaxation; mood; sensory processing
substance P (SP)	peptidergic	pain signaling, other functions
neuropeptide Y (NPY)	peptidergic	appetite control
opioids (Enk)	peptidergic	pain modulation; satiety
adenosine triphosphate (ATP)	purinergic	many functions

Receptors[2]

Neurotransmitter release can have two kinds of effects on the post-synaptic cell, depending on the number and type of receptors in its membrane. The first is ionotropic, in which a receptor allows ions to pass through the membrane when activated. The second is metabotropic, in which a receptor triggers internal biochemical signaling when activated.[3] Most neurotransmitters can activate specific receptors of both kinds. The post-synaptic cell can change its response by altering the number and type of receptors on the post-synaptic membrane.

Ionotropic receptors that are linked to sodium channels will excite a neuron and make it more likely to fire, whereas those linked to chloride channels tend to suppress firing by returning membrane potential to its resting state. The resulting changes in membrane potential are called post-synaptic potentials. Post-synaptic potentials can be either excitatory (EPSP) or inhibitory (IPSP), depending on whether they make the neuron more or less likely to fire. Dozens of different synapses may be activated in close proximity to each other, opposing or reinforcing each other's effects in a mix of competing influences. These effects spread passively across the membrane, but the effects diminish with distance. Whether or not the charge on the cell membrane reaches the threshold and triggers an action potential is determined by the mix of influences felt at the trigger zone, usually located where the axon leaves the cell body at the axon hillock. This means that the influence of a synapse varies according to how close it is to the axon hillock. For example, the chandelier cells of the cerebral cortex activate ionotropic chloride channels located directly on the axon hillock, so their effect is to instantly cancel whatever activity might be pushing the postsynaptic neuron towards triggering an action potential. This enables them to function as a silencer of axonal outputs.[4]

Metabotropic receptors activate internal signaling molecules to produce a wide variety of effects on postsynaptic cells. These effects are typically longer-lasting than the brief gushes of ions through ionotropic receptors. Some metabotropic receptors cause opening or closing of ion

---

[2] The term *receptor* can refer either to cells that are sensitive to stimuli such as light and touch, or to these membrane protein structures that respond to neurotransmitters.

[3] These functional classes are not clearly separated. Some ionotropic receptors let calcium into the cell, which can also affect biochemical pathways by activating calcium-dependent enzymes. Conversely, some metabotropic receptors activate or close membrane ion channels via internal second messengers.

[4] The distance between a synapse and the hillock is usually a good predictor of influence, but the shape and thickness of dendrites also influences the spread of synaptic influence. In addition, actively propogated potential spikes may be triggered, exerting far more influence than simple synaptic potentials.

channels by means of internal messenger molecules, while others change the expression of genes and the production of proteins. Another type of metabotropic receptor releases calcium from intracellular stores, sparking oscillating cycles of enzyme activity or causing changes to the structure of the synapse itself. It is not unusual for a synapse to have both ionotropic and metabotropic receptors.

Gap junctions

Gap junctions are small openings in the cell membrane that connect neurons together, allowing ions and therefore electric currents to pass from one neuron to another with no physical barrier. Because of this, changes in membrane potential in one cell can have an instant influence on connected cells.

In some cases, the gap junction system is combined with chemical synaptic communication. Certain large networks of inhibitory neurons in the brain are gap-junction coupled, and also make inhibitory synapses on each other. In this situation, an action potential in one neuron causes a

Gene expression

Gene expression involves the decoding of genetic information in DNA to make some sort of non-DNA product, usually a protein. The proteins are molecules that build the various parts of the body. Gene expression in individual neurons can be detected by a number of techniques, including in situ hybridization.

Neuroglia

The neuroglia are supporting cells in the nervous system. The main types are astrocytes, oligodendrocytes, microglia, and Schwann cells.

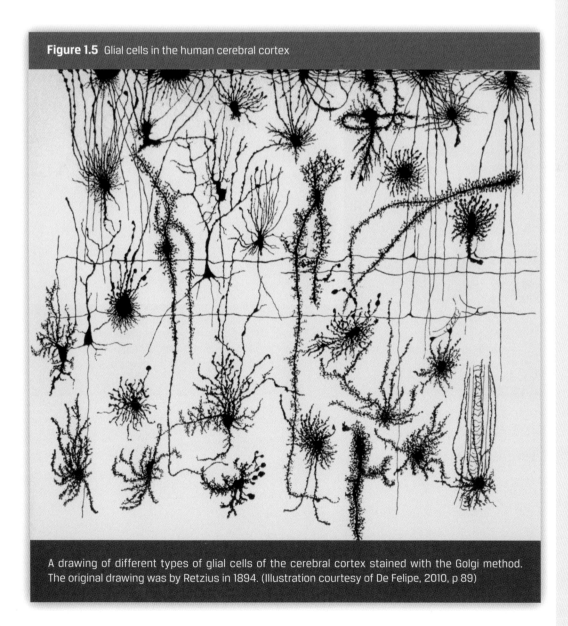

Figure 1.5 Glial cells in the human cerebral cortex

A drawing of different types of glial cells of the cerebral cortex stained with the Golgi method. The original drawing was by Retzius in 1894. (Illustration courtesy of De Felipe, 2010, p 89)

brief simultaneous electrical excitation in neurons linked to it by gap junctions, which is followed by a synaptic inhibition. The timing of these two actions causes specific frequencies of activity to spread more easily between neurons, and the network may end up firing in synchrony. This large-scale coordination is thought to be important in the activity of large areas of cerebral cortex, and may even contribute to the neuronal basis of attention.

Glia

Glia were long regarded as the simple glue filling in the space between neurons, but they are now recognized as important functional partners to neurons. While neurons tend to communicate with fast, specific signals, glial activity in the same region may take the form of gradual shifts in electrical excitability. Glial impulses spread through the tissue at a slower rate and alter neuronal activity, neurotransmitter kinetics, metabolic activity, blood flow, and the extracellular environment. Since glia and neurons are in contact across almost every part of their membranes, they interact and influence each other in a tight, reciprocal relationship, which is not completely understood.

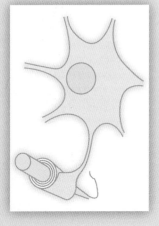

Figure 1.6
Oligodendrocyte
A diagram of a process of an oligodendrocyte which has wrapped around an axon to form a myelin sheath.

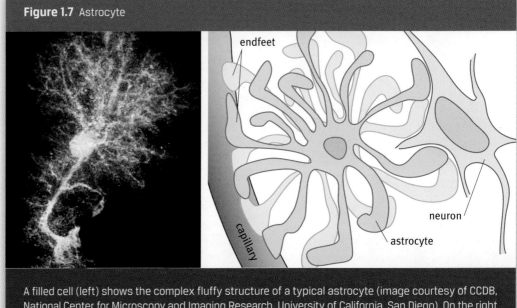

Figure 1.7 Astrocyte

A filled cell (left) shows the complex fluffy structure of a typical astrocyte (image courtesy of CCDB, National Center for Microscopy and Imaging Research, University of California, San Diego). On the right is a diagram illustrating the buffering role of an astrocyte that lies between a blood capillary and a neuron.

Astrocytes

The principal glial type in the central nervous system is the astrocyte–a general-purpose cell whose functions include buffering the cellular environment, tending synapses, metabolic regulation, signaling and communication, and governing capillary flow and traffic between neural tissue and the bloodstream. Individual astrocytes fill non-overlapping domains with fine, fluffy processes surrounding virtually every square micron of neuronal membranes and capillary walls. A single astrocyte interacts with dozens of neurons, regulating their environment and metabolism, and communicating back and forth to regulate synapses and neuron excitability. Astrocytes also provide neurons with pre-processed food to meet their energy demands. This relationship, combined with their ability to alter the size of brain capillaries, allows them to change blood flow according to neuronal energy use. Apart from meeting metabolic demands efficiently, these changes are the basis of functional magnetic resonance imaging (See Chapter 11).

Oligodendrocytes

These cells manufacture myelin, the multiple layers of fatty cell membrane that encircle many axons. The myelin sheath is not continuous; there is a small gap between the sheath made by one oligodendrocyte and the next. This gap is called a node. Myelin sheaths speed up signaling along the axon by allowing action potentials to jump from one node to the next. Not a lot is known about the other functions of oligodendrocytes. However, they have recently been shown to respond to synaptic events with post-synaptic potentials and action-potential-like spikes, further blurring the line between the functions of neurons and glia.

Schwann cells

Outside the central nervous system, the myelinating role of oligodendrocytes is taken by Schwann cells. Schwann cells manufacture myelin that encircles the axons of peripheral nerves. As in the central nervous system, the myelin sheath of peripheral nerve axons is not continuous; there are gaps between the sheath made by one Schwann cell and the next. Besides their role in myelination, Schwann cells have similar functions to the astrocytes of the central nervous system. Schwann cells are important in the maintenance of the extracellular environment around neural tissue, and in the uptake and processing of neurotransmitters around synapses, particularly at the neuro-muscular junction.

Microglia

Microglia are specialized macrophage cells. Macrophages are immune system cells that are found in blood and other tissues of the body. When macrophages enter the brain, they change into microglia. Microglia assume a stationary sentry role, exploring the tissue around them with fine processes that continuously extend and retract. If the brain is damaged or infected, microglia are activated and can resume their mobile form, which is able to engulf and destroy pathogens, foreign material, and necrotic tissue. They also signal to other cells of the immune system for assistance.

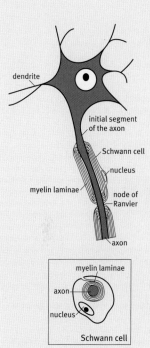

Figure 1.8
Myelinated axon

A diagram of the first part of a myelinated axon. The initial unmyelinated segment of the axon is called the axon hillock. In this case two myelin condensations are seen, the first of which is associated with the nucleus of a Schwann cell. The myelin in each condensation is produced by cell membrane extensions of a Schwann cell. Between adjacent condensations of myelin is a short length of axon not covered by myelin. This is called a node of Ranvier. (Adapted from Brodal, 2004, p 20)

Oligodendrocyte precursor cells–polydendrocytes

In addition to the mature glia described above, the central nervous system contains precursor cells which can be transformed into oligodendrocytes, astrocytes and possibly even neurons. These cells are also called polydendrocytes. They are capable of communicating and signaling in neuron-like ways, and play an important role in regeneration after injury.

Ependymal cells

The fluid spaces of the brain (the ventricles) are lined by ependymal cells. In some places the ependymal cells form specialized clusters with blood vessels called choroid plexuses. The choroid plexuses secrete cerebrospinal fluid into the ventricles. Some ependymal cells have the ability to divide and produce new neurons throughout the life of the organism. Steady streams of developing neurons from the ependymal layer migrate to the olfactory bulb and the dentate gyrus of the hippocampus, where they replace old neurons, which may have died or become dysfunctional.

Cerebrospinal fluid

Cerebrospinal fluid is a clear watery fluid that fills the ventricles of the brain and the space surrounding the brain.

How many glial cells?

Old estimates put the number of glial cells in the brain as at least 10 times the number of neurons, but recent work shows that the number of glial cells is about the same as the number of neurons.

Computers and nervous systems

Biologists and computer scientists have often compared nervous systems with the functions and properties of computers. This analogy has been useful for some understanding some neural functions, and has suggested new paradigms and algorithms to assist the design of computer architectures and applications. However, these comparisons are limited by the fact that computers and nervous systems operate in fundamentally different ways.

Computer models of neural activity have taken two basic forms–simulations of the activity of actual neurons, and simulations of neural networks, using neuron-like connection patterns and rules derived from physiology, to solve computational problems. The latter has allowed new approaches to computationally difficult problems, but is nearly always implemented using standard computer hardware and programming languages, which make it no more similar to neural activity than conventional computer programs. The modeling of actual neural activity is really a scaled-up and more sophisticated version of the neural network paradigm, except that individual neurons are modeled in much greater detail. Neither approach captures much of the biological character of neural tissue, particularly the ability of neural systems to continually modify themselves by changing gene expression and other variables. The models allow simple ideas about neural activity patterns to be explored, but they are not realistic enough to be very useful.

It could be argued that by describing nervous systems in terms of computational activity, these models are missing the point entirely, since they exhibit no evidence of awareness or consciousness.

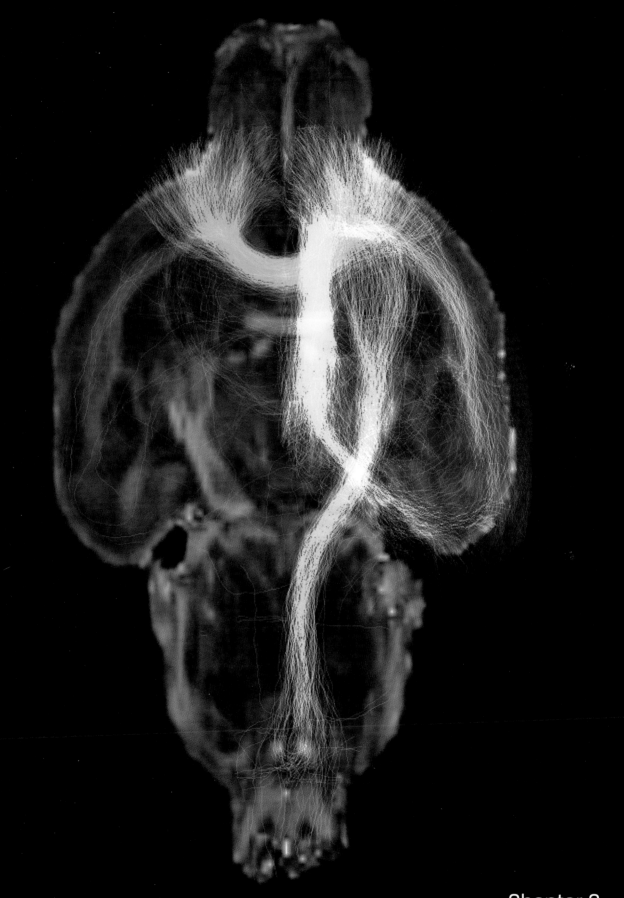

Chapter 2
Central nervous system basics—the brain and spinal cord

2 Central nervous system basics – the brain and spinal cord

In this chapter:

Why study animal brains? 12

The main parts of the brain 14

External features of the brain 15

The language of brain anatomy 18

The spinal cord 19

Learning your way around the brain involves three separate but overlapping processes. You have to start by learning some basic geography (topography)–what are the main components of the brain and how do you recognize them? Next you need to get an idea of the location of the major functional areas of the brain, such as the motor and sensory areas, since geography by itself can be meaningless. Finally, in learning the geography and finding the main functional parts, you will have to learn a new language–the language of brain anatomy. This chapter will cover the basic geography of the brain, a few of the important functional components, followed by an outline of the basic elements of the language of brain anatomy.

Why study animal brains?

Fortunately for the study of brain function, most areas of the brain are basically the same in all mammals. The outstanding exception is the huge development of the cerebral cortex in humans, which is shared by only a few mammals. Apart from the cerebral cortex, all mammals share even fine details. This means that mice and rats can be studied to work out details of connections and function, except for special cortical functions that can only be researched in humans and other primates.

It is not surprising that brains of animals are so similar. In the process of vertebrate evolution, the motor systems, sensory systems, and behaviors essential for survival appeared hundreds of millions of years ago. We share these basic survival systems with other mammals.

Comparing different brains is in some ways like comparing automobiles. Even though cars look and perform very differently from each other, they have most of the same basic parts. When you compare a huge Cadillac with a tiny Fiat 500, you find that both have four wheels, brakes, a petrol engine with cylinders and spark plugs, a steering wheel, indicator lights, and so on. The differences are in size, the level of sophistication of some of the systems, and the fact that some expensive cars have special features such as navigation computers that other cars do not. The same concept applies to brains. The rat brain and the human brain have all the basic features of a typical mammalian brain, but the human brain has a massively developed cerebral cortex with many special features that are not found at all in the brains of rats and mice.

Despite the special features of the human brain, many important studies on the cause of human brain disease have been based on work on the rat brain. For example, much of what we know about Parkinson's disease was revealed by experimental studies on rat brains. The rat brain model for Parkinson's disease has enabled researchers to test new drug treatments for Parkinson's disease. Since the discovery of gene targeting in mice, the mouse brain has been the center of work on the effect of gene mutations.

This book will therefore commence with an introduction to the features of the brains of the two most widely used experimental animals–the rat and the mouse. The main reason for postponing the detailed study of the human brain is the sheer complexity of human brain anatomy. Because some parts of the human brain have grown so large, it is much easier to understand the relationship between the main parts of the brain by studying the rodent brain first. Once the main systems of the brain have been understood, the special human features can be investigated.

Figure 2.1 Comparison of rat and human brain

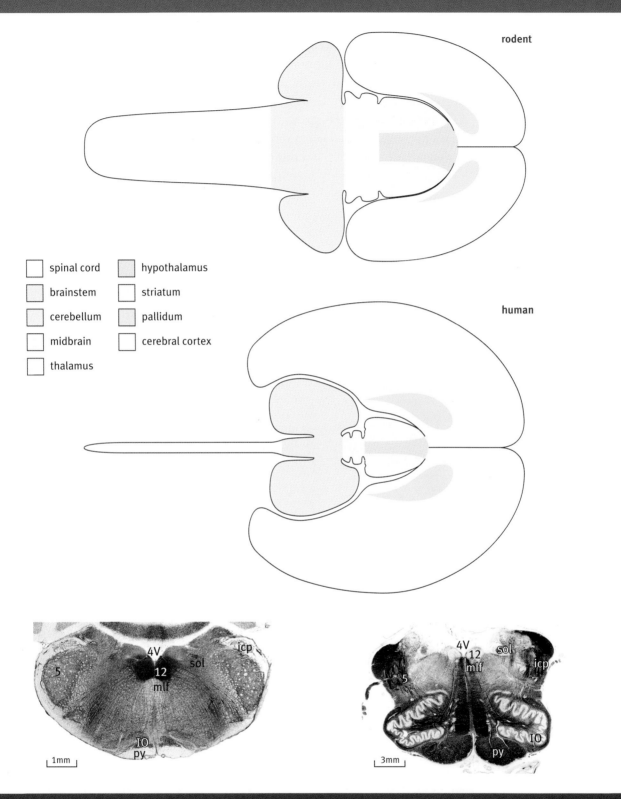

A comparison between rodent and human nervous systems, using flat-map representations pioneered by Professor Larry Swanson. Although both contain the same functional divisions, the proportions differ greatly, especially the cerebral cortex. Inset are corresponding cross sections through the hindbrain: the rat is stained for acetylcholinesterase and the human is stained for myelin. Although the proportions are different, they have the same structures (py=pyramidal tract; sol=solitary tract; 5=sensory trigeminal nucleus; 12=hypoglossal nucleus; icp=inferior cerebellar peduncle; 4V=fourth ventricle).

The main parts of the brain

All mammalian brains start off as a series of three swellings in the embryonic neural tube. The three swellings are called the brain vesicles. The brain vesicles progressively enlarge to form the forebrain, midbrain, and hindbrain vesicles. The forebrain vesicle forms a midline part, the diencephalon and hypothalamus. The lateral part forms the cerebral vesicle. Each cerebral vesicle develops into the cerebral cortex and deep cerebral structures. The midbrain vesicle forms the first part of the brainstem and four swellings on its surface, the superior and inferior colliculi. The hindbrain vesicle forms the last part of the brainstem and the cerebellum.

Neural tube

The neural tube is an embryonic cylinder of cells that develops into the central nervous system.

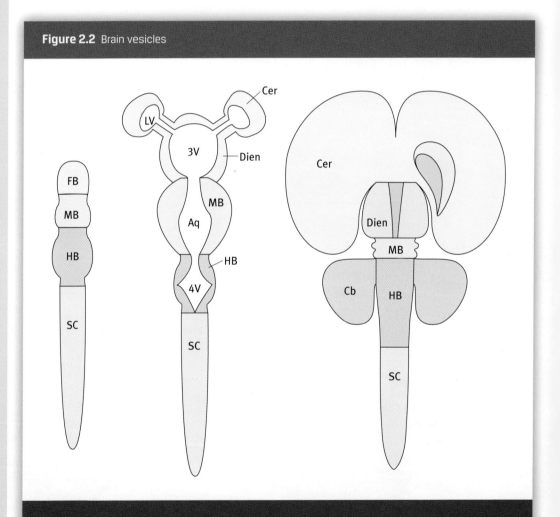

Figure 2.2 Brain vesicles

Three stages of brain development are shown in this diagram. In the early developing neural tube (left) three swellings appear. They are the forebrain, midbrain, and hindbrain vesicles. The forebrain vesicle develops a pair of lateral extensions, the cerebral vesicles (Cer), which eventually form the expanded cerebral hemispheres. The remaining midline part of the forebrain vesicle forms the diencephalic vesicle (Dien) and the hypothalamus (not shown here). Each of the vesicles surrounds a fluid filled chamber; these chambers are called the lateral ventricle (LV), the third ventricle (3V), the aqueduct (Aq), and the fourth ventricle (4V). The cerebellum (Cb) grows out from the developing hindbrain. (Flat map adapted from Swanson, 2004).

External features of the brain

The first stage of understanding brain anatomy is to be able to recognize the main regions of the brain. To assist your appreciation of this section on external anatomy of the brain, you may wish to access a series of high quality photographic images of a rat brain that is available at www.apuche.org/OIA/Anatomical-Page-03.htm.

The dorsolateral aspect of a rodent brain

We will start with a picture of the rat brain as viewed from side after it has been taken out of the skull. The front of the brain is on the right-hand side and the back end, which connects to the spinal cord, is on the left. The front end of the brain is called the forebrain and we can see two main parts, the small olfactory bulb on the far right, and the large cerebral hemisphere next to it. The olfactory bulb is the main receiving center for the sense of smell and it is quite large in most mammals, except in humans, where it is relatively tiny.

On the left-hand side of this view of the brain, we can see the cerebellum and the brainstem. The part of the brainstem under the cerebellum is called the hindbrain. Connecting the hindbrain to the forebrain is the midbrain. The midbrain is not very large but it contains vital centers for processing visual and auditory (hearing) information and for controlling basic movement patterns. The characteristic feature of the midbrain is the group of four little bumps on the upper

Olfactory functions in mammals

Many mammals have a keen sense of smell and have large olfactory bulbs. In the human brain the olfactory bulb is relatively small. Most mammals also have an additional olfactory system for detecting sexual signaling pheromones. Humans do not possess this special olfactory system.

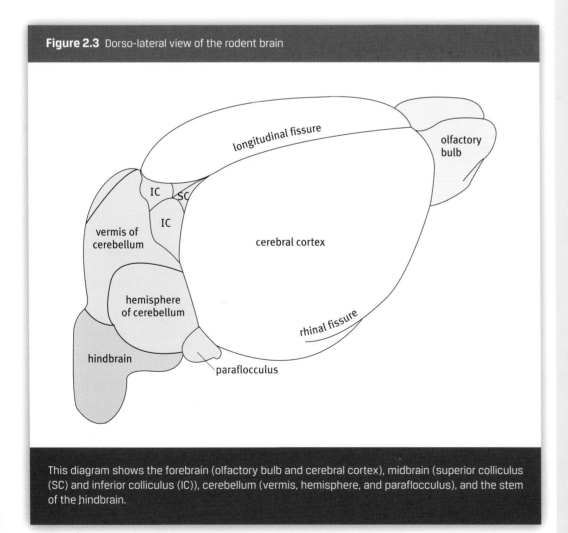

Figure 2.3 Dorso-lateral view of the rodent brain

This diagram shows the forebrain (olfactory bulb and cerebral cortex), midbrain (superior colliculus (SC) and inferior colliculus (IC)), cerebellum (vermis, hemisphere, and paraflocculus), and the stem of the hindbrain.

surface. Each one of these is called a colliculus (which is Latin for a little hill). The two colliculi closest to the cerebral hemispheres are called the superior colliculi, and the two closest to the cerebellum are called the inferior colliculi.

The surface of the cerebellum is marked by dozens of tiny folds and looks a bit like a small cauliflower. Because of its many folds it will be easy to recognize in cross-sections. Under the cerebellum is the part of the brainstem that connects the midbrain to the spinal cord. This part of the brainstem is often called the medulla oblongata, but it is best referred to as the hindbrain.

The hindbrain is partly a corridor for bundles of connections linking the brain to the spinal cord, but it is also the home of the centers that control our breathing, swallowing and state of consciousness. If the hindbrain is squashed or injured in any way, the individual often dies. The reason that boxers become unconscious (and sometimes die) after a powerful punch to the jaw is that the hindbrain is slammed against the bones of the skull.

Rostral

Rostral is a directional term, meaning towards the nose, away from the tail.

Caudal

Caudal is a directional term, meaning towards the tail.

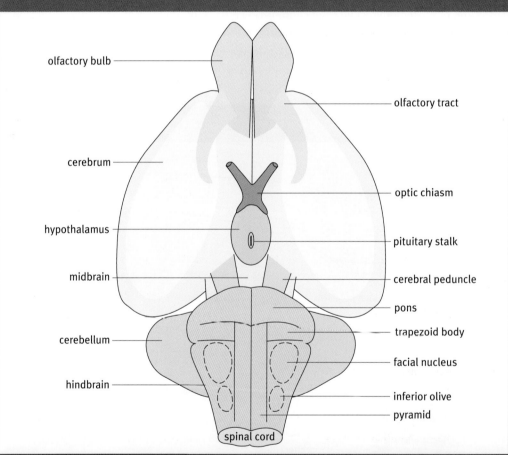

Figure 2.4 Ventral view of the rodent brain

This is a diagram of the ventral surface of the mouse brain. The hypothalamus is in the centre of the diagram and the stalk of the pituitary is attached to the hypothalamus. The rostral end of the hypothalamus is marked by the optic chiasm. The only part of the midbrain that can be seen is the cerebral peduncle. The hindbrain consists of a central stem and the cerebellum. Two transverse fiber bundles are seen in the rostral hindbrain- the pons and the trapezoid body. The pyramids are two longitudinal fiber bundles that run from the pons to the junction of the spinal cord. The position of the facial nucleus and the inferior olive are shown lateral to each pyramid. (Adapted from Farris and Griffith, 1949, p 40)

The ventral aspect of a rodent brain

The next view of the brain is from underneath. Here the two olfactory bulbs at the top of Figure 2.4 and the large cerebral hemispheres are easy to identify. Between the two cerebral hemispheres is the diencephalon, which connects the midbrain to the rest of the forebrain. The diencephalon is hidden by the hypothalamus in Figure 2.4.

The main feature on the under-surface of the hypothalamus is the stalk that connects it to the pituitary gland. The pituitary is the master gland of the endocrine (hormone) system, but it in turn is under the control of the hypothalamus. In front of the pituitary stalk are the two optic nerves, which form a cross called the optic chiasm. Behind the pituitary stalk is a pair of small bumps called the mammillary bodies. The mammillary bodies contain some of the nerve cell groups that make up the hypothalamus.

On each side of the hypothalamus is the part of the cerebral hemisphere known as the temporal lobe. The bulge at the back of the temporal lobe is the amygdala, a complex group of nerve cell clusters buried inside the temporal lobe. The amygdala has an important role in responding to threats to survival.

Next we return to the midbrain and hindbrain. The under-surface of the midbrain is located just behind the mammillary bodies. In the hindbrain, the most prominent feature is a band of fibers running from side to side, called the pons. These crossing fibers are part of a system which connects the forebrain to the cerebellum. A large nerve emerges on each side of the pons. This is the trigeminal nerve, the nerve that carries sensation from the face to the hindbrain.

If you look closely at the space between the pons and the mammillary bodies, you can see a pair of thick fiber bundles. These are the cerebral peduncles, which are the main fiber bundles connecting the forebrain with the brainstem, cerebellum and spinal cord. The word peduncle means a stalk. The space between the two cerebral peduncles is called the interpeduncular fossa (fossa is Latin for a hole or a ditch). A pair of nerves emerges from the fossa; these are the oculomotor nerves, the main nerves that control eye movement (Figure 4.5).

To finish our tour of the under-surface of the brain, we will look at the structures on the lower (caudal side) of the pons. The first thing to notice is a pair of fiber bundles close to the midline, which connect the forebrain with the spinal cord. They are called the pyramids, even though they do not really look like pyramids in the rat. At the side of each pyramid is a series of bulges made by cell groups inside the hindbrain. The one closest to the pons is called the superior olive (an auditory center, here covered by the trapezoid body), and the next one caudal to it is the facial nucleus (the cell group that controls the facial muscles). The last one, which is harder to make out, is the inferior olive. The inferior olive is connected to the cerebellum and is unrelated to the superior olive.

A number of pairs of nerves emerge from the hindbrain (Figure 4.5). They include nerves related to hearing, balance, and swallowing, but we will not try to identify them all at this stage.

A mid-sagittal section of a rodent brain

The next view (Figure 2.5) is that of the brain when it is cut into two halves down the midline. This cut is called a mid-sagittal section. When the left and right halves are completely separated we can look at the middle of the brain from front to back. This view allows us to see an important range of structures that are hidden or only partly visible from the outside.

The sagittal section of the brain shows us all the main subdivisions of the brain in a single view.

Diencephalon

The diencephalon is one of the major subdivisions of the forebrain, and lies between the two cerebral hemispheres. The parts of the diencephalon are the pretectum (prosomere 1), the thalamus (prosomere 2), and the prethalamus (prosomere 3). The hypothalamus was formerly included in the diencephalon but gene expression analysis shows that it is a distinct entity.

Section planes

The most common section planes used in neuroanatomy are coronal, sagittal, and horizontal. Coronal sections, also called transverse sections, are cut at right angles to the long axis of the nervous system, the neuraxis. They show neuroanatomy from a "face on" perspective. Sagittal sections split the brain lengthways, from rostral to caudal, at right angles to the horizontal plane. Horizontal sections cut the brain lengthways in the horizontal plane.

On the right-hand side you can see the hindbrain and cerebellum. On the left-hand side you can see the main parts of the forebrain—the olfactory bulb, the cerebrum, the thalamus, and the hypothalamus. In between the forebrain and the brainstem is the midbrain, which can be recognized by the pair of rounded bumps on its upper surface (the superior colliculus and the inferior colliculus).

Attached to the hypothalamus and thalamus are two large glandular structures, the pituitary gland and the pineal gland. The pituitary gland is attached to the under surface of the hypothalamus and the pineal gland is attached to the upper surface of the thalamus just where it joins the midbrain.

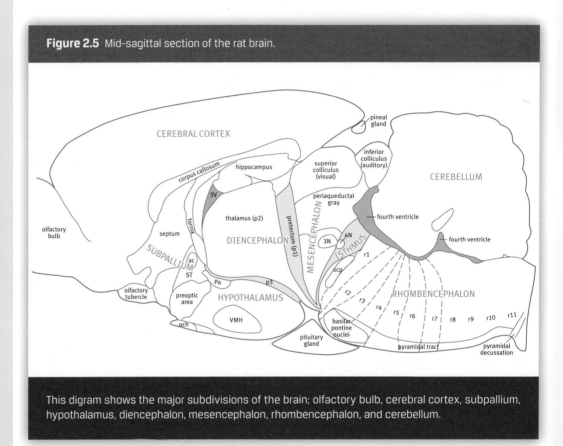

Figure 2.5 Mid-sagittal section of the rat brain.

This digram shows the major subdivisions of the brain; olfactory bulb, cerebral cortex, subpallium, hypothalamus, diencephalon, mesencephalon, rhombencephalon, and cerebellum.

Cranial nerves

The twelve pairs of nerves that are joined directly to the brain are called the cranial nerves. Some are connected to the sense organs of the head (eye, nose, ear), and some supply the muscles, glands, and skin of the head and neck.

The language of brain anatomy

When you want to describe the position of brain structure A in relation to structure B, you could say "A is next to B" or "A is above B," but these general descriptions are not precise enough to be really useful. Because of this, a special descriptive language has been developed. The main terms are in pairs—medial and lateral, superior and inferior, and anterior and posterior. The terms rostral and caudal are sometimes used instead of the more traditional terms anterior and posterior. Rostral means closer to the front end of the neural axis, whereas caudal means closer to the tail end of the neural axis. The use of the terms rostral and caudal is particularly useful in reference to the human brain because of the ninety-degree bend at the level of the midbrain.

The brain is made up of groups of neurons and bundles of axons. Neurons may be grouped in a cluster, which is called a nucleus, or arranged in a sheet (layer or lamina), and called an area. In the midbrain and hindbrain, most neurons are arranged in nuclei, but in the cerebral cortex and cerebellar cortex they are arranged in laminae.

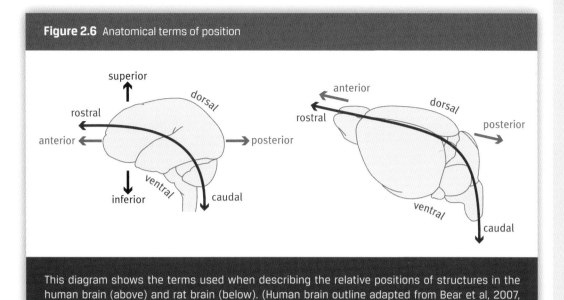

Figure 2.6 Anatomical terms of position

This diagram shows the terms used when describing the relative positions of structures in the human brain (above) and rat brain (below). (Human brain outline adapted from Bear et al, 2007, p 402)

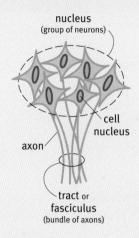

Figure 2.7
A group of neurons and a bundle of axons

A group of neurons is called a nucleus, but note that this term is also applied to the nuclear region of a cell. A bundle of axons is called a tract or fasciculus.

Bundles of nerve fibers (axons) are given various different names. When they form an elongated bundle, they can be called a tract, a fasciculus, or a faniculus. The name of a tract usually describes its beginning and end. For example, the corticospinal tract starts in the cerebral cortex and ends in the spinal cord. Large tracts that are oval-shaped in cross section may be called a lemniscus. The huge tract that connects the cerebrum to the brainstem and thalamus has a special name, the internal capsule. The name relates to the way the fiber sheet forms a capsule or coating around the thalamus.

Tracts that cross the midline are called either commissures or decussations. Examples are the anterior commissure of the cerebrum and the corticospinal (pyramidal) decussation at the caudal end of the brainstem. The special name chiasma (chiasm) is given to the crossing of the optic nerves; the word chiasma refers to the Greek letter chi. The largest of all commissures joins the two cerebral hemispheres. It contains millions of axons and is called the corpus callosum.

The spinal cord

The spinal cord is the part of the central nervous system that controls the voluntary muscles of the limbs and trunk, and which receives sensory information from these same regions. It also controls most of the viscera and blood vessels of the thorax, abdomen, and pelvis.

The spinal cord and its meningeal coverings (dura mater, arachnoid mater, and pia mater) lie within the vertebral canal, a hollow space that runs inside the spinal vertebrae. Early in development, the spinal cord is the same length as the vertebral column, but the vertebral column grows faster than the spinal cord, and in adult humans the cord extends only two-thirds of the way down the vertebral canal. Below this the vertebral canal contains only the spinal nerve roots and meninges.

The human spinal cord has eight cervical segments, twelve thoracic segments, five lumbar segments, five sacral segments, and one coccygeal segment, making a total of 31 segments. The spinal cord is enlarged in the segments connected to the nerves of the upper limb (C5 to T1) and

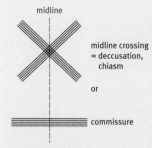

Figure 2.8
Commissures and decussations

Crossing fibers are called commissures when the crossing is symmetrical and decussations when the crossing is asymmetrical.

Figure 2.9 White and gray matter in the spinal cord and the cerebrum

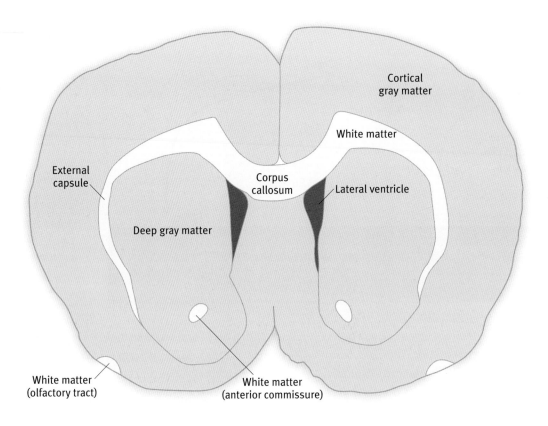

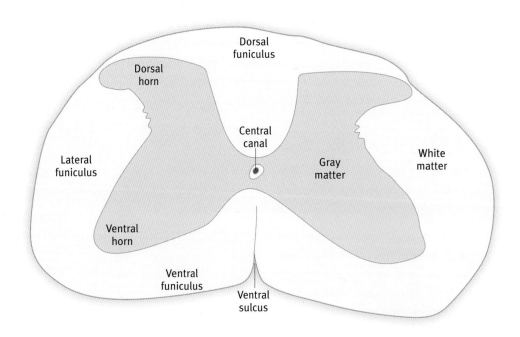

In the spinal cord the gray matter forms a central core surrounded by white matter. In the cerebrum this arrangement is reversed because the gray matter of the cortex covers white and gray structures in the center.

the segments connected to the nerves of the lower limb (L2 to S2). These enlargements are due to the large numbers of motor neurons present at these levels.

Spinal nerves

Pairs of spinal nerves arise from the spinal cord and leave the vertebral column through the intervertebral foramina. Each spinal nerve is formed by a number of dorsal and ventral rootlets. The dorsal rootlets are bundled together to form the dorsal root of a spinal nerve and the ventral rootlets form the ventral root. The dorsal roots contain only sensory fibers, while the ventral roots contain only motor fibers. Each dorsal root has an ovoid swelling called a spinal (dorsal root) ganglion. Spinal ganglia consist of neurons with bifurcated axons, one part connecting to the periphery and the other traveling into the spinal cord.

Gray matter of the spinal cord

The spinal cord gray matter is made up of neuronal cell bodies and neuropil (the space between neurons that contains, glia, dendrites, and axons). The gray matter can be macroscopically divided into dorsal and ventral horns with an intermediate region between them.

A transverse section of spinal cord shows that the gray matter is arranged in the form of a butterfly or the letter H, depending on the level. The cross bar of the H, called the commissural gray matter, encloses the central canal. The dorsally projecting arms of the gray matter are called the dorsal horns, and the ventrally projecting arms are called the ventral horns. The dorsal horn of gray matter reaches out to connect with the surface of the spinal cord at the point where the

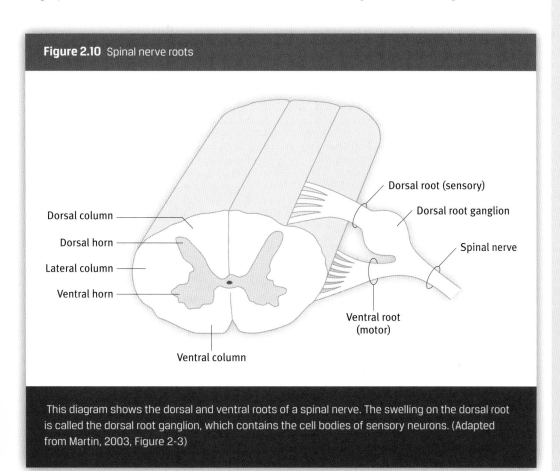

Figure 2.10 Spinal nerve roots

This diagram shows the dorsal and ventral roots of a spinal nerve. The swelling on the dorsal root is called the dorsal root ganglion, which contains the cell bodies of sensory neurons. (Adapted from Martin, 2003, Figure 2-3)

Dorsal root ganglion

The dorsal root ganglion is a swelling on the dorsal root of a spinal nerve that contains the cell bodies of the primary sensory neurons of that nerve.

Neuropil

The term neuropil describes the space between neuron cell bodies as seen in histological sections. The neuropil is full of axons, dendrites, synapses, and glial cells, but these structures are not seen in standard histological preparations such as Nissl-stained sections.

dorsal nerve rootlets enter the spinal cord. The central region, which connects the dorsal and ventral horns, is called the intermediate gray matter. In the thoracic spinal cord, as well as the first two lumbar segments, and in the upper sacral segments, there is a small lateral projection of the intermediate gray matter called the intermediolateral column. The intermediolateral column contains the cells of origin of the autonomic nervous system. The thoracic intermediolateral column contains preganglionic sympathetic neurons, and the sacral intermediolateral column contains preganglionic parasympathetic neurons.

White matter of the spinal cord

A layer of white matter surrounds the gray matter except for where the dorsal horn touches the surface of the spinal cord. The white matter consists mostly of longitudinally running axons, many of which form distinct bundles called tracts. The horns of gray matter divide the white matter into three columns, dorsal, lateral and ventral. The two dorsal columns are located between the two dorsal horns of gray matter. The boundary between the lateral and ventral horns is not distinct. It is generally taken to be in line with the emerging axons of the most lateral motor neurons. While the two dorsal columns lie side by side with a common medial border, the two ventral columns are separated by a deep fissure, the ventral median fissure.

Autonomic nervous system

The autonomic nervous system is the part of the nervous system that controls the activity of the visceral organs, blood vessels, and glands.

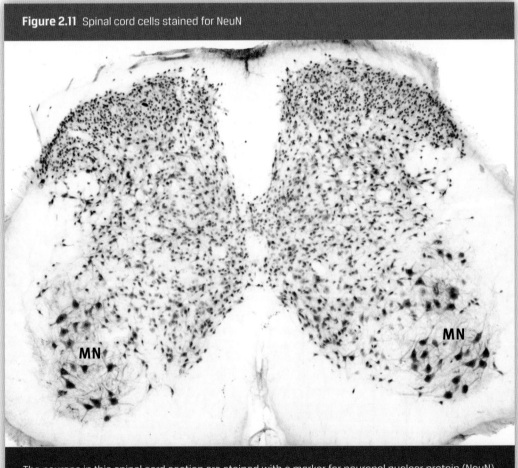

Figure 2.11 Spinal cord cells stained for NeuN

The neurons in this spinal cord section are stained with a marker for neuronal nuclear protein (NeuN). The large multipolar neurons (MN) in the ventral horn are motor neurons which supply limb muscles. Each cluster of motor neurons relates to a particular muscle group.

Meninges, ventricles, and the circulation of cerebrospinal fluid

The brain and spinal cord are suspended in a clear liquid called the cerebrospinal fluid (CSF). The CSF is contained by a tough membrane inside the skull and vertebral canal, called the dura mater. Thin elastic fibers, called the arachnoid, connect the dura mater to the pia mater, a thin membrane clinging to the surface of the brain and spinal cord. Because the brain is so soft and easily damaged, the cerebrospinal fluid plays an important protective role by acting as a shock absorber. The cerebrospinal fluid is manufactured inside the brain by networks of small arteries called choroid plexuses. The choroid plexuses are bound to adjacent ependymal cells, which line the inner surface of the ventricles. The cerebrospinal fluid flows from the lateral ventricles to the third ventricle, which is a narrow slit between the two halves of the hypothalamus. The fluid then travels along a narrow tube, called the aqueduct of the midbrain. The aqueduct drains into the fourth ventricle, which lies between the cerebellum and the brainstem. The fluid escapes through three holes in the roof of the fourth ventricle to fill the subarachnoid space. Cerebrospinal fluid is eventually reabsorbed into the blood circulation by specialized structures in the walls of large veins that line the inside of the skull.

If the circulation of cerebrospinal fluid is impeded, the results can be serious. For example, a prenatal condition that blocks the flow of cerebrospinal fluid will cause fluid to accumulate in the cerebral hemispheres. After a time, the pressure causes the cerebral ventricles to expand and cells in the cerebral cortex are killed off. This condition is known as hydrocephalus (commonly known as 'water on the brain').

Meninges

The meninges are three membrane layers which cover the brain and spinal cord. The layers are the dura, arachnoid, and pia mater. The dura is a thick fibrous layer inside the surrounding bones, and the pia clings to the surface of the brain and spinal cord. Between them is the delicate arachnoid mater, which is connected to the pia by fine weblike strands. The cerebrospinal fluid lies between the arachnoid and the pia.

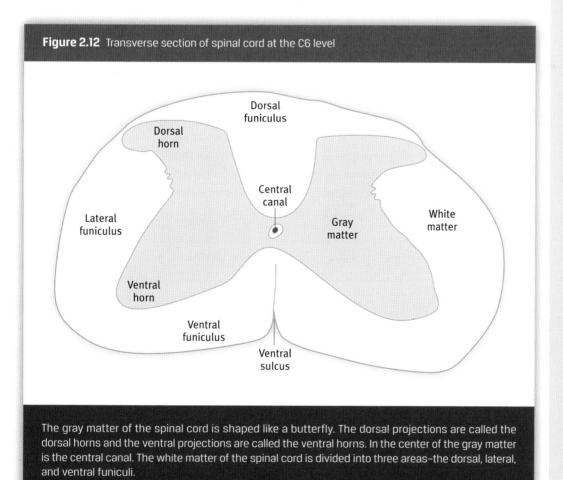

Figure 2.12 Transverse section of spinal cord at the C6 level

The gray matter of the spinal cord is shaped like a butterfly. The dorsal projections are called the dorsal horns and the ventral projections are called the ventral horns. In the center of the gray matter is the central canal. The white matter of the spinal cord is divided into three areas–the dorsal, lateral, and ventral funiculi.

The blood-brain barrier

The blood-brain barrier is a unique vascular arrangement of the brain that prevents certain molecules passing from the blood to the extracellular fluid of the brain. Unlike other capillary systems in the body, which allow most molecules to pass through without restriction, the traffic into the extracellular fluid (ECF) of the brain is tightly regulated. This is important in controlling the delicate balance for neuronal excitability as well as protecting the brain from toxins and infectious particles.

The blood-brain barrier is formed by the specialized membranes of the endothelial cells of the brain capillaries. The endothelial cells of brain capillaries are joined together by tight junctions, which block substances from passing though the wall of the capillary. In addition, brain endothelial cells have few endocytotic vesicles compared to other endothelial cells, and this limits transcellular transport into the ECF. As a result, molecules that are not transported must pass through the plasma membrane of the endothelial cells themselves to enter the ECF. Water-soluble substances cannot pass through the plasma membrane, but small lipid-soluble molecules may diffuse through the endothelial cell to the ECF. This has important implications for drug treatment as only lipophilic drugs can reach the brain in adequate concentrations.

The physical barrier created by the endothelial cells is supplemented by cellular transport mechanisms that control the passage of substances from the blood to the ECF. Brain endothelial cells have specialized transporter molecules and cell membrane pumps that keep unwanted molecules out of brain circulation. For example the p-glycoprotein transporter protein on the endothelial membrane is important in expelling small lipid-soluble molecules that have entered the cell through passive diffusion. The brain endothelial cells also have specialized transporters to pump in substances wanted by the cells of the brain. The brain has a high glucose requirement, and the provision of glucose is assisted by a specialized glucose transporter that is present on the brain endothelium. Glucose in the blood binds to the GLUT-1 transporter and is actively pumped into the brain extracellular fluid.

The blood-brain barrier is essential in protecting the brain from infection and toxins. Most bacteria and their harmful toxins are too large to pass through the tight junctions between the endothelial cells. However, inflammatory processes can alter the permeability of the blood-brain barrier, and in some cases this allows pathogens directly into the central nervous system. Infections such as this can be life threatening.

Certain areas of the central nervous system are not protected by the blood-brain barrier. These areas are referred to as the circumventricular organs and include the area postrema in the floor of the fourth ventricle, the subfornical organ and the vascular organ of the lamina terminalis in the walls of the third ventricle, the median eminence of the hypothalamus, the choroid plexus, the pineal gland, and the posterior lobe of the pituitary gland.

Circumventricular organs

Circumventricular organs are located in the walls of the ventricles of the brain, so the neurons there can directly sense the levels of various chemicals in the bloodstream. Some of the circumventricular organs, such as the pineal gland and the median eminence, can also secrete substances directly into the blood stream. Some scientists argue that the posterior pituitary should be included as one of the circumventricular organs because it contains neuronal processes that secrete oxytocin and vasopressin directly into the bloodstream.

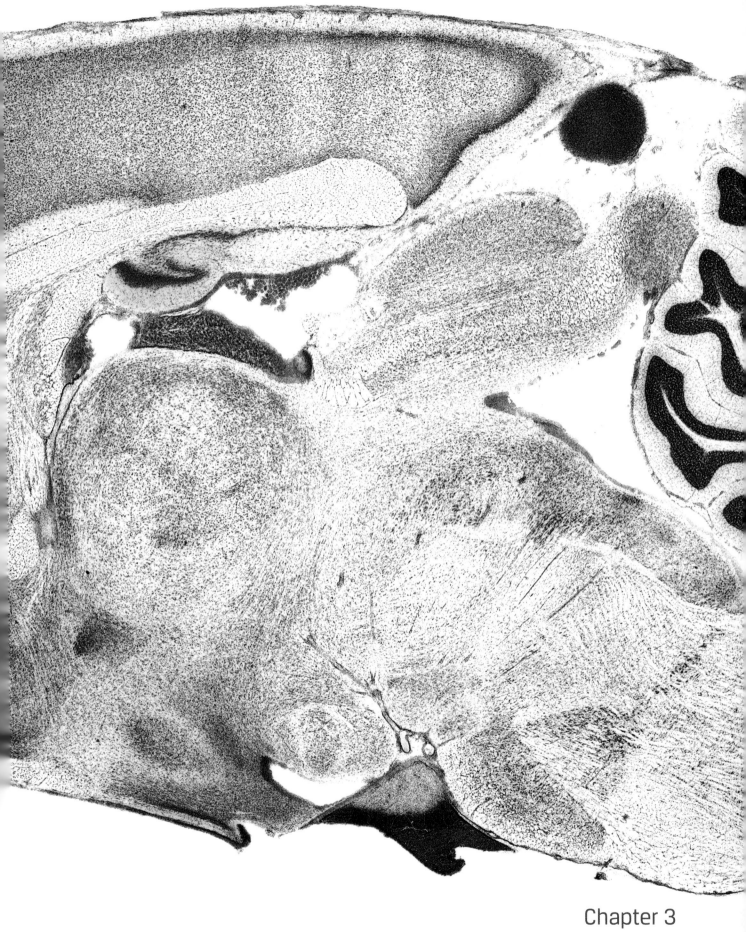

Chapter 3
A map of the brain

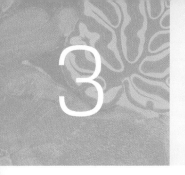

3 A map of the brain

In this chapter:

Mini-atlas of the rat brain   26

Mini-atlas of the rat brain

When the brain is cut into thin sections, these sections can be stained to show the nuclei and fiber tracts. A series of sections forms an atlas, which shows us the position of the main structures inside the brain. This section of the book is a set of simplified drawings of brain sections (a mini-atlas) to assist those who are less familiar with the detailed anatomy of parts of the brain. The mini-atlas contains twelve diagrams of coronal sections from one end of the rat brain to the other. The major regions of the brain and some important nuclei and tracts are labeled. To assist your understanding of these simplified sections, each drawing is accompanied by a short description of the main features. Following the twelve coronal diagrams are diagrams of two sagittal sections of the brain. These latter sections show some areas of the brain to advantage, and are particularly useful for understanding the relationships between diencephalic, midbrain, and hindbrain regions.

The brain is made up of three main parts, the forebrain, the midbrain (also called mesencephalon), and the hindbrain. The forebrain is divided into the cerebral cortex, the subpallium, the hypothalamus, and the diencephalon. Each of these main subdivisions is labeled in Figure 3.1b. The midbrain joins the hindbrain to the forebrain. The hindbrain has three major parts, the isthmus, the rhombencephalon, and the cerebellum. The end of the rhombencephalon joins with the spinal cord.

Figure 3.1

The upper image (a) is photograph of a Nissl-stained sagittal section of a rat brain. The section is close to the midline. The dark folds of the cerebellum are seen on the right, and the dark blue stripe on the far left is a part of the olfactory bulb. The dark diamond-shaped patch ventral to the middle of the brain is the pituitary gland.

The lower image (b) is a color-coded diagram of the photograph seen in (a). The major parts of the brain are labeled. Each of the structures shown here will be discussed in the remaining pages of this chapter.

Figure 3.2

This diagram of a sagittal section of the rat brain shows the position of the coronal section diagrams which form an atlas series in this chapter (Figures 3.3 to 3.14). Each vertical line shows the position of the relevant coronal section. This diagram should be consulted when each of the coronal diagrams is studied.

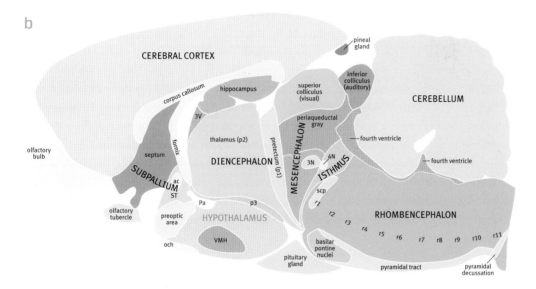

Figure 3.1
Sagittal section of the rat brain

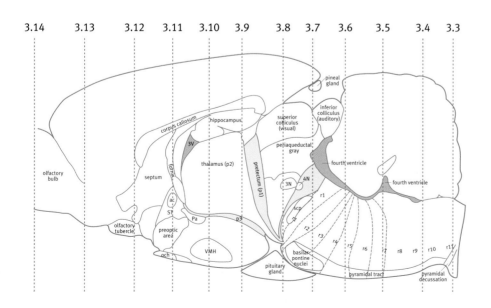

Figure 3.2
Sagittal finder diagram with coordinates

The Brain   Watson Kirkcaldie Paxinos

Figure 3.3
Coronal section at the junction of the hindbrain and spinal cord

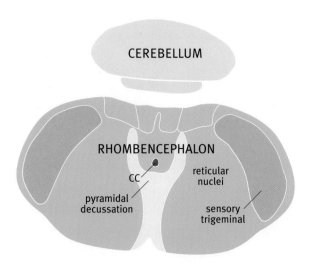

The junction of the hindbrain and spinal cord

This section shows the brainstem at the point where it joins the spinal cord. Sitting dorsal to the hindbrain is the caudal part of the cerebellum. In the center of the hindbrain is the prominent crossing of the pyramidal tract (pyramidal decussation). The fibers in the pyramidal tract arise in the cerebral cortex and are here seen crossing to the opposite side of the spinal cord. The large trigeminal nuclei, which receive touch, pain, and temperature sensations from the face, are found in the lateral part of the hindbrain. This part of the trigeminal sensory complex is called the spinal trigeminal nucleus because it extends into the cervical spinal cord. Most of the remainder of the area of the section is occupied by the reticular nuclei, which extend the whole length of the hindbrain. The part of the cerebellum seen here is the caudal end of the vermis, the midline part of the cerebellum. In this figure, the ventricular system of the brain is represented by the small central canal (CC), which continues down through the center of the spinal cord. The central canal extends to the lower end of the spinal cord, but it is a blind canal from which the cerebrospinal fluid cannot escape.

Hindbrain versus brainstem

The term hindbrain means the entire region between the midbrain and the spinal cord. It therefore includes the central core of the hindbrain and the cerebellum. However, the term hindbrain is often used to refer only to the central core, without reference to the cerebellum. The term brainstem is used to refer to the midbrain and the central core of the hindbrain, but excluding the cerebellum.

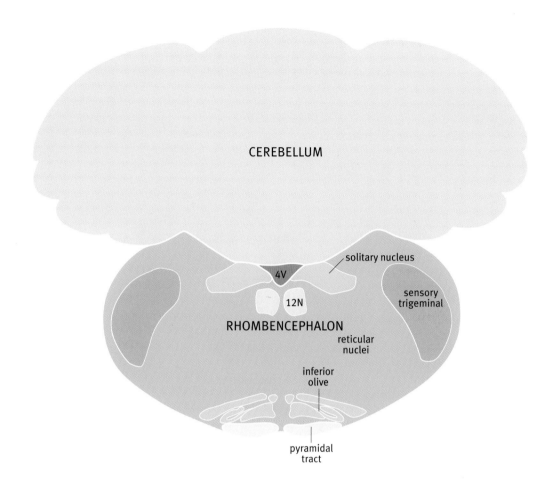

Figure 3.4
Coronal section of the hindbrain at the level of the inferior olive

The hindbrain at the level of the inferior olive

This section shows the brainstem below and the cerebellum above with the fourth ventricle in between. This part of the hindbrain that lies under the cerebellum is the rhombencephalon.

The large trigeminal nuclei are found in the lateral part of the hindbrain. This part of the trigeminal sensory complex is called the spinal trigeminal nucleus. The two hypoglossal nuclei (which control the tongue) are next to the midline under the fourth ventricle. Lateral to the hypoglossal nucleus is a cluster of cell groups called the solitary nucleus. This cluster receives taste sensation and sensory information from internal organs such as the stomach and the lung. Two large fiber bundles, the pyramids, lie on either side of the midline on the ventral margin of the hindbrain. The fibers in the pyramids arise in the cerebral cortex and travel down to cross to the opposite side of the spinal cord. Next to each pyramid is the inferior olive, which is functionally connected with the cerebellum. Most of the remainder of the area of this section is occupied by the reticular nuclei, which extend for the whole length of the hindbrain. They are involved in basic sensory and motor functions.

The fourth ventricle contains cerebrospinal fluid (CSF), which has flowed down from its origin in the lateral (cerebral) ventricles through the third ventricle and aqueduct to reach the hindbrain. The CSF escapes from the roof of the fourth ventricle to fill the subarachnoid space.

The cerebellum is a large structure concerned with coordination of movement. It consists of an outer layer of cerebellar cortex and a core of white matter (fibers).

Rhombencephalon

The rhombencephalon, a section of the neural tube, is the embryonic precursor of the hindbrain.

Pyramids

The pyramids are large bundles of corticospinal axons on the ventral surface of the hindbrain next to the midline.

Figure 3.5
Coronal section of the hindbrain at the level of the facial nucleus

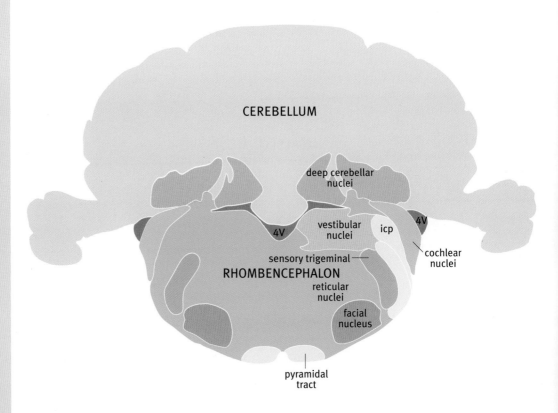

The hindbrain at the level of the facial nucleus

This section shows the hindbrain below and the cerebellum above with the fourth ventricle in between. The part of the hindbrain that lies under the cerebellum is called the rhombencephalon.

The large trigeminal nuclei are found in the lateral part of the hindbrain at this level. This part of the trigeminal sensory complex is called the spinal trigeminal nucleus because it extends into the cervical spinal cord. Two large fiber bundles, the pyramids, lie on either side of the midline on the ventral margin of the hindbrain. The pyramids contain the pyramidal (corticospinal) tracts. Between each pyramid and the trigeminal nucleus is a large group of motor neurons that supplies the muscles of facial expression on that side, called the facial nucleus. The vestibular nuclei lie at the junction of cerebellum and hindbrain at the side of the fourth ventricle. They receive information from the position sense organs of the inner ear. At the lateral edge of the hindbrain under the cerebellum are the cochlear nuclei, which receive auditory sensations from the inner ear. Most of the remainder of the section is occupied by the reticular nuclei, which extend for the whole length of the hindbrain.

The cerebellum consists of an outer layer of cerebellar cortex and a core of white matter (fibers). At this level, some cell groups can be seen in the deep cerebellar white matter dorsal to the ventricle and the vestibular nuclei. These cell groups are the cerebellar nuclei. They receive input from the cerebellar cortex and send fibers to the brainstem and thalamus.

The fourth ventricle has lateral extensions at this level. They can be seen next to the cochlear nuclei.

The fourth ventricle

The fourth ventricle is a tent-shaped space located between the core of the hindbrain and the cerebellum. The roof of the fourth ventricle has three holes that allow cerebrospinal fluid to escape into the subarachnoid space. If these holes are blocked, the fluid pressure in the ventricles rises, causing expansion that can kill brain cells. This is called hydrocephalus.

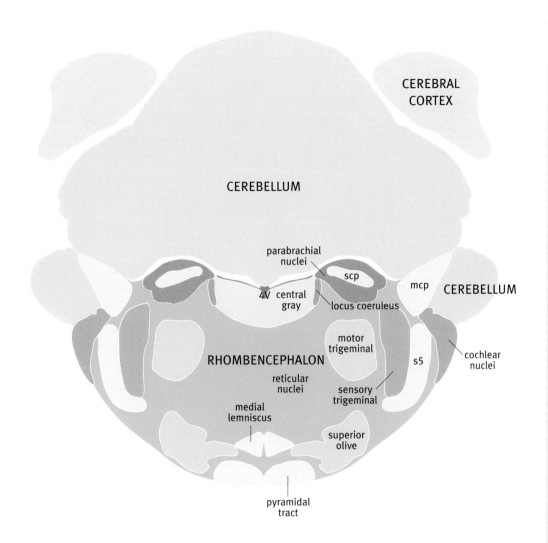

Figure 3.6
Coronal section of the hindbrain at the level of the upper end of the fourth ventricle

The hindbrain at the level of the upper end of the fourth ventricle

This section shows the hindbrain below and the rostral part of the cerebellum above. Dorsal and lateral to the cerebellum is the caudal end of the cerebral cortex. The rostral end of the fourth ventricle lies between the brainstem and the cerebellum. The part of the brainstem seen here is the rostral end of the hindbrain. The large trigeminal nuclei are found in the lateral part of the hindbrain. This part of the trigeminal complex is called the principal sensory trigeminal nucleus. It receives touch sensation from the face. Medial to this is the motor trigeminal nucleus, which controls the chewing (masticatory) muscles. Two large fiber bundles, the pyramids, lie next to the midline on the ventral margin of the hindbrain. Just above each pyramid is the medial lemniscus, a fiber bundle carrying touch information from the body and face to the thalamus. The thalamus then sends this information to the cerebral cortex. Lateral to the trigeminal nuclei are the cochlear nuclei. Between the pyramid and the trigeminal nucleus at this level is the superior olive, a collection of nuclei that process and analyze auditory information. Above the superior olive is an area occupied by the reticular nuclei. The cerebellum is dorsal to the core of the hindbrain. The cerebellum is joined to the hindbrain by two fiber bundles, the middle cerebellar peduncle (mcp) and the superior cerebellar peduncle (scp). Between the cerebellum and the core of the hindbrain is the fourth ventricle (4V). The fourth ventricle contains cerebrospinal fluid (CSF) that has flowed down from its origin in the lateral (cerebral) ventricles.

Lemniscus

A lemniscus is a large bundle of axons. The major somatosensory tract in the brainstem is called the medial lemniscus.

The locus coeruleus

The locus coeruleus is a small cell group located in the rostral part of the floor of the fourth ventricle. It can be seen as blue spot in fresh brains. It contains almost all the noradrenaline neurons of the brain. The locus coeruleus neurons send their axons to reach the whole of the forebrain, brainstem, cerebellum and spinal cord. These neurons are vital to keeping the brain awake and alert, and when they are shut down, the brain goes to sleep.

Figure 3.7
Coronal section of the brainstem at the level of the midbrain

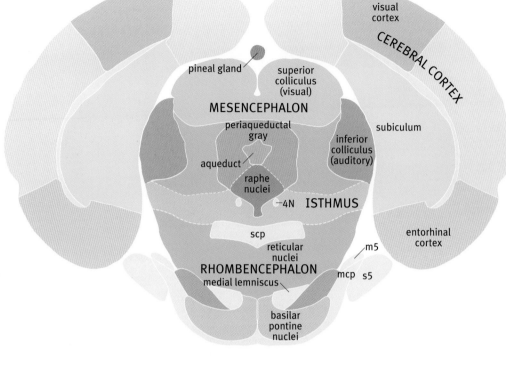

The brainstem at the level of the midbrain

This section shows the brainstem in the center covered above and to the side by the bulging caudal end of the cerebral cortex. Because the neuraxis is flexed in this region, the top half of the brainstem section includes the midbrain, whereas the bottom half includes the rostral parts of the hindbrain, such as the pons. The pons is made up of the basilar pontine nuclei and the fibers that connect to them. The fibers that arise from the pontine nuclei cross to reach the opposite side of the cerebellum (pons is Latin for bridge). The fibers of the pyramidal tract are embedded in the pontine nuclei. Just above the pons is the medial lemniscus, a fiber bundle carrying touch information from the body and face to the thalamus. Sitting just lateral to the pons is the sensory trigeminal nerve (s5), a large bundle of fibers that carries sensory information from the skin of the face to the trigeminal sensory nuclei.

In the center of the midbrain is the aqueduct, a small canal along which cerebrospinal fluid (CSF) flows from the ventricles of the forebrain to the fourth ventricle of the hindbrain. The aqueduct is surrounded by a thick layer of cells called the periaqueductal gray. Below the periaqueductal gray is the small motor neuron cluster called the trochlear nucleus (the nucleus of the fourth cranial nerve). Each trochlear nucleus supplies just one of the six muscles that move each eye. Above the periaqueductal gray is the pair of superior colliculi. Each superior colliculus receives visual sensory information from the opposite eye. Lateral and ventral to the superior colliculus is the inferior colliculus, which receives auditory information from the superior olive.

The part of the cerebrum seen here is the occipital pole, which is occupied mainly by the visual cortex. The entorhinal cortex occupies the ventral zone and the subiculum (part of the hippocampal complex) occupies the medial zone of cortex. The pineal gland is attached to the dorsal surface of the thalamus. It manufactures the hormone melatonin, which has a role in the sleep-wake cycle.

Brainstem

Brainstem is a collective term used to describe the midbrain and the non-cerebellar part of the hindbrain. It is a useful term, but from a technical point of view it is better to refer to the midbrain and hindbrain individually wherever possible.

The periaqueductal gray and pain

The periaqueductal gray (PAG) is a thick layer of neurons around the aqueduct of the midbrain. It plays a vital role in the response of the brain to severe pain. Some PAG neurons contain endorphins, which have an action like opium or morphine. Drugs like morphine may act on the PAG to relieve the suffering associated with severe pain. Some PAG neurons connect with the serotonin neurons of the raphe nuclei.

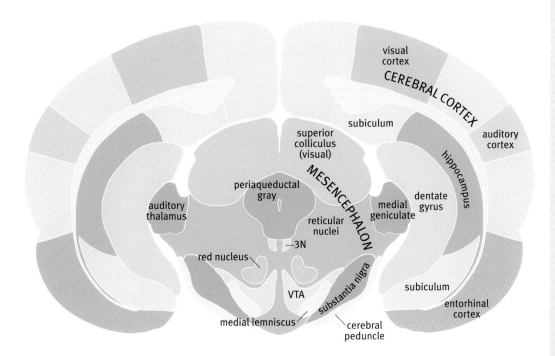

Figure 3.8
Coronal section at the rostral end of the midbrain

The rostral end of the midbrain

This section shows the midbrain in the center covered above and to the side by the bulging caudal end of the cerebral cortex. The superior colliculus occupies the dorsal part of the midbrain, but the lateral part contains the most caudal part of the thalamus–the medial geniculate nucleus, which is the auditory thalamic nucleus.

At this level, the pyramidal (corticospinal) fibers and other descending fibers form a bundle called the cerebral peduncle on the ventral aspect of the midbrain. Coating the inside (medial) border of the fibers of the cerebral peduncle is the substantia nigra, a large cell group that is important for motor control. Fibers from the substantia nigra project to the deep nuclei of the cerebrum, especially the striatum. Just above the medial end of the substantia nigra is the medial lemniscus, a fiber bundle carrying touch information from the body and face to the thalamus. On the medial side of the medial lemniscus is another motor area, the ventral tegmental area (VTA).

Surrounding the aqueduct is a thick layer of neurons called the periaqueductal gray. Below the periaqueductal gray is a prominent pair of motor nuclei called the oculomotor nuclei. Each oculomotor nucleus (3N) supplies four of the six muscles that move each eye. In addition, the oculomotor nerve supplies the main muscle of the upper eyelid and the muscles of the pupil and lens. Between the periaqueductal gray and the medial lemniscus is the red nucleus. This large group of cells gives rise to a fiber bundle that travels down the spinal cord, called the rubrospinal tract. Between the two cerebral peduncles at the ventral margin is the interpeduncular nucleus. The area between the red nucleus and the superior colliculus is filled with reticular nuclei. Attached to the lateral edge of the midbrain is the medial geniculate nucleus of the thalamus.

The part of the cerebrum seen here contains the visual cortex above and the auditory cortex laterally. The ventral part of the cortex is occupied by the entorhinal cortex, and the medial part is occupied by the hippocampal structures (hippocampus, dentate gyrus, and subiculum).

Superior colliculus

The superior colliculus is a raised area of the dorsal surface of the midbrain involved in visual reflexes and attention.

Substantia nigra

The substantia nigra is a cell group in the ventral part of the brainstem that is important in motor control. Many cells in the substantia nigra contain dopamine, and give rise to a dopaminergic pathway which projects to the striatum.

Entorhinal cortex

The entorhinal cortex is an area of cortex in the temporal lobe close to the hippocampus that encodes a spatial map of the external surroundings for use by the hippocampus.

Figure 3.9
Coronal section of the forebrain at the level of the caudal end of the thalamus

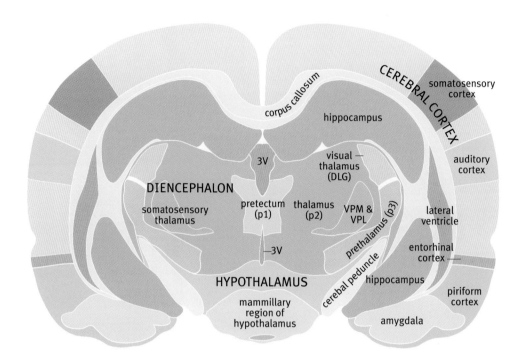

The forebrain at the level of the caudal end of the thalamus

This section shows the thalamus and hypothalamus in the center covered above and to the side by the cerebrum. In the midline is the third ventricle, which links the lateral ventricles to the aqueduct.

The thalamus (prosomere 2 of the diencephalon) lies dorsal to the hypothalamus. The thalamus receives input from touch, auditory, and visual pathways, and sends fibers to the cerebral cortex. The part of the thalamus that receives sensory information from the eyes is called the dorsal lateral geniculate nucleus (DLG). The dorsal lateral geniculate connects with the visual cortex of the cerebrum. The somatosensory (touch sensation) nuclei of the thalamus (VPL and VPM) lie ventral to the DLG. On the medial edge of thalamus is part of the pretectal area (prosomere 1 of the diencephalon), and the lateral edge is the prethalamus (prosomere 3).

The hypothalamus is located ventral to the thalamus. At this level of the hypothalamus, there is a pair of bulges called the mammillary bodies. At the lateral edge of the hypothalamus is the cerebral peduncle. The cerebral peduncle contains the pyramidal (corticospinal) fibers as well as the fibers traveling from the cortex to the pons.

The part of cerebral cortex seen here is about midway between occipital and frontal poles of the cerebrum. The most obvious cortical areas at this level are the somatosensory cortex above, the auditory cortex laterally, and the piriform (olfactory) cortex below. Above the piriform cortex is the rostral end of the entorhinal cortex. Medial to the piriform cortex is a large group of nuclei called the amygdala. The amygdala is associated with emotional responses and a number of survival behaviors Medial to the lateral ventricle is the hippocampus, an easily recognizable part of cerebral cortex involved with memory registration. The corpus callosum is a large sheet of fibers connecting the left and right sides of the cerebrum.

Prosomere

A prosomere is a segmental component of the developing diencephalon. The three prosomeres are the pretectum (prosomere 1), the thalamus (prosomere 2), and the prethalamus (prosomere 3).

Thalamus

The thalamus is the major part of the diencephalon; it receives input from touch, auditory, and visual pathways (among others), and sends fibers to the cerebral cortex.

Hypothalamus

The hypothalamus is a major subdivision of the forebrain located ventral to the thalamus. It is responsible for the initiation of survival behaviors and controls the autonomic and neuroendocrine systems.

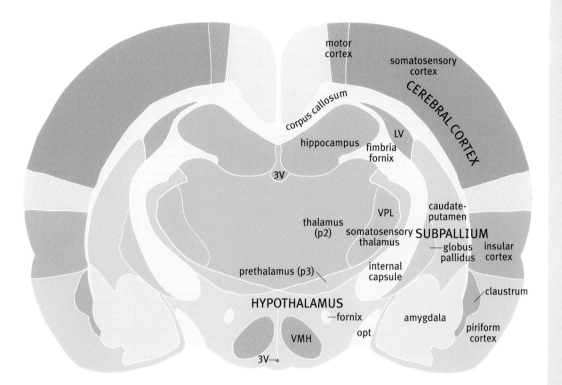

Figure 3.10
Coronal section of the forebrain at the level of the internal capsule

The forebrain at the level of the internal capsule

This section shows the thalamus and hypothalamus in the center covered above and to the side by the cerebrum. In the midline is the third ventricle, a cavity that links the lateral ventricles to the aqueduct. Separating the thalamus from the cerebrum is the internal capsule, a very large sheet of fibers that connects the cerebral cortex with the thalamus, brainstem and spinal cord. The caudal end of the internal capsule forms the cerebral peduncle.

The section cuts through the somatosensory (touch sensation) cortex and the caudal end of the motor cortex. Below the somatosensory cortex is the insular cortex (taste and visceral sensation) and the piriform cortex. Medial to the piriform cortex is a large group of nuclei called the amygdala.

The hippocampus is an easily recognizable part of cerebral cortex involved with memory registration. In this section, the hippocampus is seen between the corpus callosum and the thalamus. The thin layer of white matter lateral to the ventricle and the caudate-putamen is called the external capsule which is continuous with the corpus callosum, the major commissure of the forebrain.

Between the internal capsule and the external capsule are members of the lateral group of deep cerebral nuclei, the caudate-putamen and the globus pallidus. These nuclei are involved in motor control, particularly semi-automatic movements and locomotion.

The thalamus (prosomere 2) receives input from the somatosensory (touch), auditory, and visual pathways, and sends projections to almost all the cortical areas. In this section, the somatosensory nucleus (VPL) is a half-moon shaped structure at the lateral edge of the thalamus.

The hypothalamus lies ventral to the thalamus, and between the thalamus and hypothalamus is the prethalamus (prosomere 3). The ventromedial hypothalamic nucleus (VMH) is a prominent landmark in the hypothalamus. Between the hypothalamus and the amygdala is the optic tract (opt), a bundle of fibers running from the optic chiasm to the dorsal lateral geniculate nucleus of the thalamus.

Insula

The insula is the area of cerebral cortex that receives taste sensation and visceral sensation from the intestines. In humans this area is an island of cortex hidden in the depths of the lateral fissure.

Hippocampus

The hippocampus is an area of the temporal lobe involved in memory.

Amygdala

The amygdala is an almond-shaped nucleus located in the anterior temporal lobe, thought to be involved in emotional responses and social behaviors.

Figure 3.11
Coronal section of the forebrain at the level of the anterior commissure

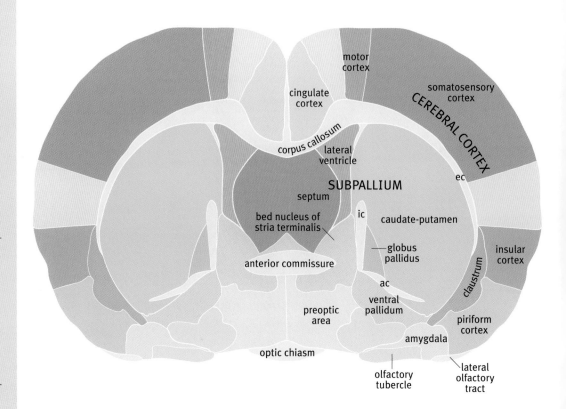

Somatosensory cortex

The somatosensory cortex is a granular area of the cortex involved with processing of touch and proprioceptive information. In humans this is found in the post-central gyrus.

Premotor cortex

The premotor cortex is the cortical area involved in planning of movement. It is located in the frontal lobe in front of the primary motor cortex.

Globus pallidus

The globus pallidus is the major part of the pallidum. It receives input from the striatum and projects mainly to the thalamus.

Striatum and pallidum

The largest of the deep cerebral nuclei are the striatum and pallidum. The parts of the striatum are the caudate, putamen, and accumbens nucleus. The parts of the pallidum are the globus pallidus and the ventral pallidum. Both striatum and pallidum contain GABA neurons which influence motor functions. The striatum receives major dopamine pathways from the substantia nigra and the ventral tegmental area.

The forebrain at the level of the anterior commissure

This section is taken immediately in front of the rostral pole of the thalamus. In the center of the section is the anterior commissure, with the septum above, the preoptic area below, and the bed nucleus of the stria terminalis around its lateral edge. In the midline of the preoptic area is the third ventricle, which links the lateral ventricles to the cerebral aqueduct. Immediately below the third ventricle is the optic chiasm, in which optic nerve fibers cross the midline.

The areas of cerebral cortex seen in this section are the somatosensory cortex and the motor cortex. Below the somatosensory cortex is the insular cortex (taste and visceral sensation) and the piriform cortex. On the surface of the piriform cortex is the olfactory tract, which connects the olfactory bulb to the piriform cortex. Medial to the piriform cortex is the rostral part of the amygdala. The cingulate cortex is seen on the medial side of the cerebrum above the corpus callosum.

Medial to the external capsule (ec) lie representatives of the lateral group of deep cerebral nuclei, the caudate-putamen and the globus pallidus. Along the medial border of the globus pallidus is the rostral part of the internal capsule (ic). The internal capsule is a large sheet of fibers which connects the cerebrum with the thalamus, brainstem, and spinal cord.

The thin layer of white matter lateral to the caudate-putamen is called the external capsule. The lateral ventricle lies between the septum and the caudate-putamen in this section. The anterior commissure (also labeled ac) is a large bundle of crossing fibers, which connects the olfactory bulb and parts of the cerebrum to the same areas on the opposite side.

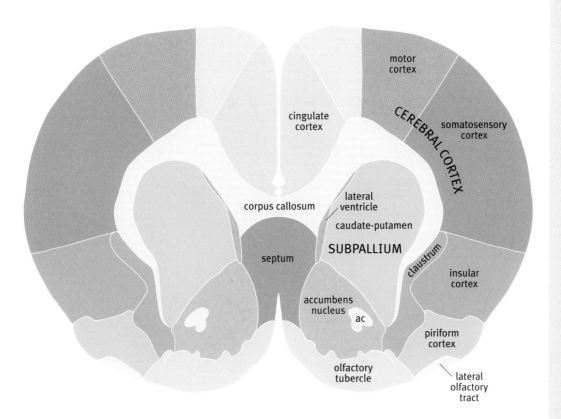

Figure 3.12
Coronal section of the forebrain at the rostral end of the corpus callosum

The forebrain at the rostral end of the corpus callosum

This section passes through the rostral end of the corpus callosum. The section cuts through the rostral end of the somatosensory cortex and the middle of the motor cortex. Medial to the motor cortex is the cingulate cortex. Below the somatosensory cortex are the insular cortex (taste and visceral sensation) and the piriform cortex. On the surface of the piriform cortex is the lateral olfactory tract which connects the olfactory bulb to the piriform cortex. Medial to the piriform cortex is the olfactory tubercle, which also receives fibers from the olfactory bulb.

The layer of white matter lateral and dorsal to the caudate-putamen is called the external capsule. Ventral to the corpus callosum is a large group of nuclei called the septum, which is the largest of the medial group of deep cerebral nuclei. Medial to the external capsule lies the lateral ventricle and the caudate-putamen. The caudate-putamen is the largest of the lateral group of deep cerebral nuclei. The lateral ventricle lies between the septum and the forceps minor (fibers of the of the corpus callosum that are traveling toward the frontal pole) medially, and the caudate-putamen laterally. Lateral to the most ventral part of the external capsule is another deep nucleus of the cerebrum called the claustrum.

The anterior limb of the anterior commissure (ac) is prominent bundle of fibers dorsal to the olfactory tubercle. Surrounding the anterior limb of the anterior commissure is the accumbens nucleus.

Corpus callosum

The corpus callosum is the major cerebral commissure. The fibers of the corpus callosum connect the two hemispheres of the cerebral cortex.

The accumbens nucleus and addiction

The accumbens nucleus is a ventral part of the striatal complex, lying in the basal forebrain close to the midline. It receives an important dopamine pathway from the ventral tegmental area. Raised dopamine levels in the accumbens are associated with feelings of satisfaction and pleasure. Most addictive drugs cause an increase in dopamine levels in the accumbens nucleus. The accumbens is thought to be connected via the thalamus to the orbitofrontal cortex, which is also associated with the perception of reward.

Figure 3.13
Coronal section of the forebrain at the level of the frontal pole of the cerebrum

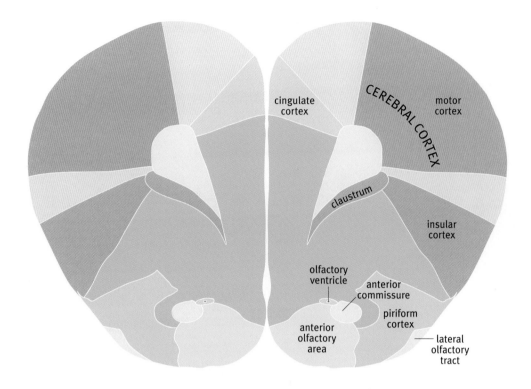

The forebrain at the level of the frontal pole of the cerebrum

The section cuts through the rostral end of the motor cortex, and below it, the motor cortex is the insular cortex and the piriform cortex. On the surface of the piriform cortex is the lateral olfactory tract, which connects the olfactory bulb to the piriform cortex. Medial to the piriform cortex is the anterior olfactory area, which also receives fibers from the olfactory bulb. Medial to the motor cortex is the cingulate cortex.

The anterior limb of the anterior commissure is embedded in the dorsal part of the anterior olfactory area. A small extension of the lateral ventricle, called the olfactory ventricle, lies adjacent to the anterior limb of the anterior commissure and extends forward into the olfactory bulb. Deep to the motor cortex is an area of white matter and below this is the claustrum.

Anterior commissure

In addition to the corpus callosum, there are three other prominent commissures in the forebrain: the anterior, hippocampal, and posterior commissures. The anterior commissure links the two olfactory bulbs and contains cortical commissural fibers from the temporal lobes.

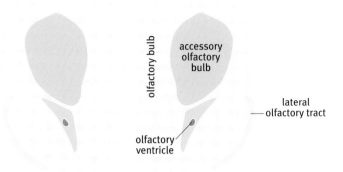

Figure 3.14
Coronal section of the olfactory bulb

The olfactory bulb

The most rostral part of the forebrain is the olfactory bulb, which receives the olfactory nerve fibers from the roof of the nose. The fibers leaving the bulb form the lateral olfactory tract. Embedded in the dorsal part of the main olfactory bulb is the accessory olfactory bulb; it receives input from the vomeronasal organ, a pheromone receptive area in the nasal cavity. These structures are not found in the human brain. A small extension of the lateral ventricle, called the olfactory ventricle, is seen in the center of the olfactory bulb.

Olfactory bulb

The olfactory bulb is a cylindrical structure lying above the roof of the nasal cavity. It is the main receiving center for the sense of smell. It is found at the rostral end of the brain in rodents, but on the ventral surface of the brain in humans.

Olfactory nerve

The olfactory nerve is the first cranial nerve; it carries the sensation of smell to the brain.

Figure 3.15
Sagittal section
-0.40mm lateral
to the midline

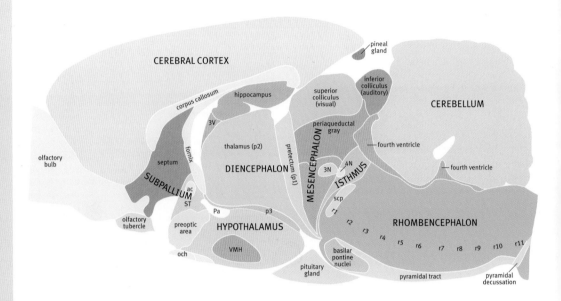

Sagittal section -0.40mm lateral to the midline

This section is close to the midline and shows all of the major parts of the brain. On the far left is the olfactory bulb and on the far right is the hindbrain. In between these two structures are the cerebrum, thalamus, and midbrain. The cerebral neocortex covers the centrally placed hippocampus, septum, and thalamus. Between the hippocampus and neocortex is the corpus callosum, a large bundle of fibers that connects the two cerebral hemispheres. Ventral to the thalamus and septum, from left to right, are the olfactory tubercle, the preoptic area, and the hypothalamus.

The thalamus is the largest part of the diencephalon. The two smaller parts of the diencephalon are the pretectum (between thalamus and midbrain), and the prethalamus (between thalamus and hypothalamus). The pituitary gland, which is attached to the hypothalamus, is seen ventral to the brain. A second gland, the pineal, is seen above the colliculi of the midbrain. The pineal has a long stalk (not seen here), which attaches it to prosomere 2 of the diencephalon.

Between the preoptic area and the septum is a collection of cell groups called the bed nucleus of the stria terminalis (ST). The stria terminalis is a large fiber bundle that starts in the amygdala and ends in the bed nucleus. The ST is also wrapped around another large fiber bundle, the anterior commissure. Three hypothalamic landmarks can be seen in this section. The optic chiasm (och) is stuck to the ventral surface of the hypothalamus, and two important hypothalamic nuclei can be seen–the paraventricular nucleus (Pa) and the ventromedial hypothalamic nucleus (VMH).

Squeezed between the pretectum and the hindbrain is the wedge-shaped midbrain (mesencephalon). The ventral compression of the midbrain is caused by the sharp flexion of the neuraxis in this region. The dorsal surface of the midbrain is marked by two bulges on each side, the superior and inferior colliculi. Deep to the colliculi is the periaqueductal gray. Embedded in the ventral part of the midbrain periaqueductal gray is the nucleus of the oculomotor nerve (3N).

The hindbrain is formed by a series of twelve embryonic segments–the isthmus and the eleven rhombomeres. The cerebellum develops from the isthmus and the first rhombomere. The last rhombomere (r11) joins the first segment of the spinal cord.

Pituitary gland

The pituitary gland is a gland attached to the base of the hypothalamus. It has two parts, the anterior and the posterior, which secrete chemicals involved in endocrine signaling systems.

Pineal gland

The pineal gland is a gland attached by a stalk to the dorsal surface of the thalamus. It releases a chemical called melatonin, which is involved in the sleep-wake cycle.

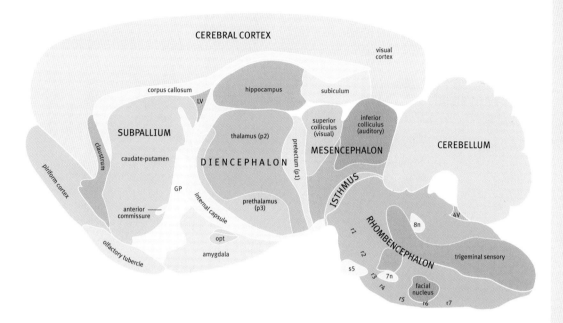

Figure 3.16
Sagittal section
−2.62mm lateral
to the midline

Sagittal section −2.62mm lateral to the midline

This section is some distance from the midline, but the major features are similar to those seen in the previous section. However, the olfactory bulb is no longer seen attached to the frontal pole of the cerebrum; the piriform cortex is there in its place. The corpus callosum forms a roof over the caudate-putamen and the hippocampus. Deep to the hippocampus is the diencephalon (pretectum, thalamus, and prethalamus). The caudal part of the cerebral neocortex overhangs the colliculi of the midbrain.

In the ventral region of the cerebrum, from left to right, we can see the piriform cortex, the olfactory tubercle, and the amygdala. The amygdala is a mass of gray matter in the temporal lobe of the cerebrum. The other matter masses of gray matter seen in this lateral region of the cerebrum are the caudate-putamen, the globus pallidus (GP), and the claustrum. The claustrum can be seen in this section pressed against the rostral part of the corpus callosum. The caudate-putamen fills the space between the corpus callosum and the olfactory tubercle. Embedded in the caudal border of the caudate-putamen are the fibers of the anterior commissure. Separating the amygdala from the diencephalon (prethalamus and thalamus) is a thick sheet of fibers, the internal capsule.

On the right of the diencephalon, caudal to the pretectum, is the midbrain (mesencephalon), which is distinctly wedge-shaped. The large dorsal surface is occupied by the superior and inferior colliculus, but the ventral surface is a small area between the pretectum and the isthmus.

The hindbrain is made up of the isthmus and the eleven rhombomeres. The cerebellum grows out of the first rhombomere and the isthmus. In this lateral region of the hindbrain, the most obvious feature is the trigeminal sensory nucleus. It extends from the second rhombomere down to the third cervical segment of the spinal cord. The sensory root of the trigeminal nerve is seen here entering the hindbrain at the level of the second rhombomere (r2). Ventral to the sensory trigeminal nucleus in the region of the sixth rhombomere (r6) is the facial nucleus, which supplies the muscles of facial expression. Although the facial nucleus is located in the sixth rhombomere (r6), the facial nerve emerges from the hindbrain at the level of the fourth rhombomere (r4).

The cerebellum consists of a folded cortex and a central core of fibers. Within this central core are the cerebellar nuclei. Between the cerebellum and the hindbrain is the fourth ventricle, one of the reservoirs of cerebrospinal fluid in the central nervous system.

Old and new cerebral cortex

The more primitive areas of cerebral cortex, such as the olfactory (piriform) cortex, have only three distinct layers. The 'new' areas of cortex (neocortex) that have evolved in mammals have six layers. More than 95% of the human cerebral cortex is neocortex.

Piriform cortex

The piriform cortex is the primary olfactory cortex. The human piriform cortex is located in the uncus.

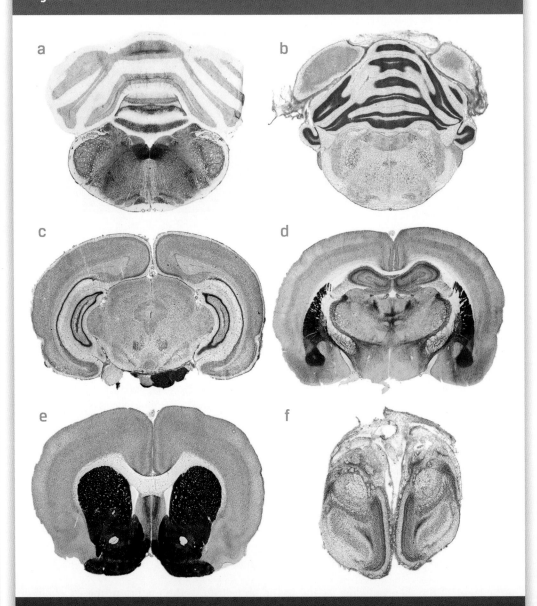

Figure 3.17 Brain sections

Photographs of coronal sections of rat brain which correspond to some of the diagrams in this Chapter. These images show that demarcating the clusters and layers of neurons that populate the brain in Nissl stained sections is not a straightforward exercise. Finding boundaries between nuclei is a complex process, requiring a great deal of information on connections and histochemistry (see Chapter 11). Sections a, d, and e are stained for acetylcholinesterase, whereas sections b, c, and f are Nissl stained. Refer to the corresponding figures for anatomical description–a:3.4; b:3.6; c:3.8; d:3.10; e:3.12; f:3.14. The images are taken from The Rat Brain in Stereotaxic Coordinates, 6th edition (Paxinos and Watson, 2007).

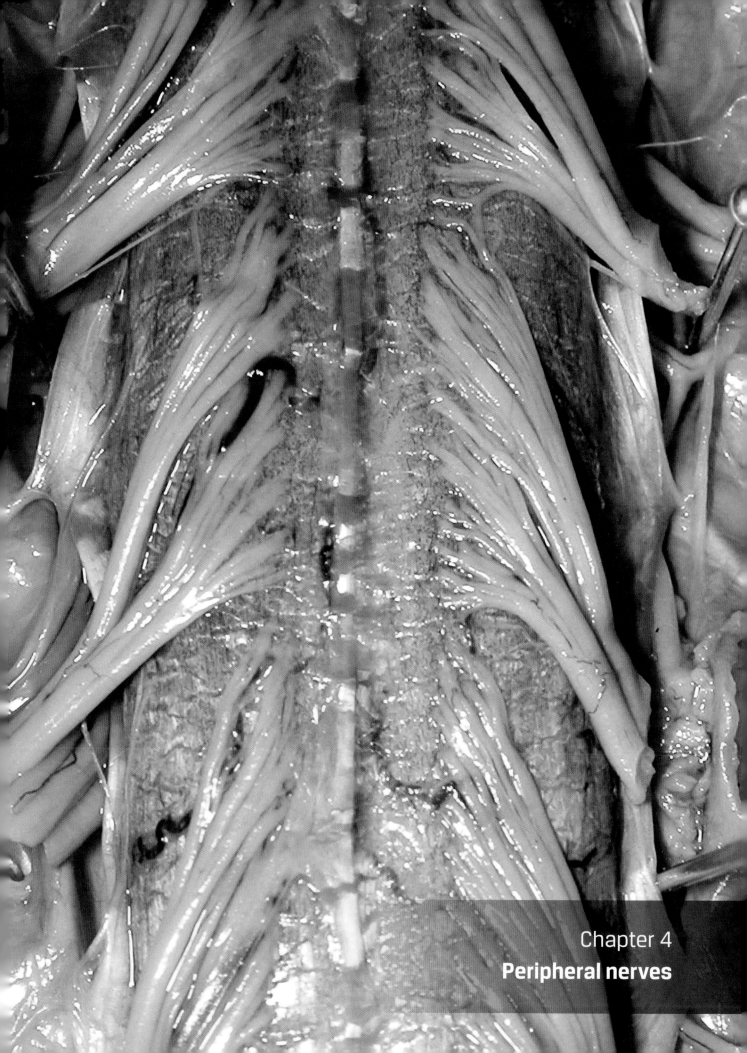

Chapter 4
Peripheral nerves

4 Peripheral nerves

In this chapter:

Motor and sensory nerves — 44

Somatic and visceral motor and sensory elements — 45

Spinal nerves — 46

Spinal nerves supplying the limbs — 47

Cranial nerves — 47

The brain and spinal cord are connected to the body by the cranial and spinal nerves. These nerves carry sensory information to the central nervous system, and they carry commands from the brain and spinal cord to activate muscles or other organs.

Motor and sensory nerves

In the early development of the neural tube there is a basic separation of motor components in the developing ventral horn (called the basal plate) from sensory components in the developing dorsal horn (called the alar plate). This is reflected in the roots of the spinal nerve. Each of the spinal nerves is split into two roots at the point where it connects to the spinal cord. The dorsal root carries sensory fibers to the dorsal horn and the ventral root is formed by motor fibers from the ventral horn.

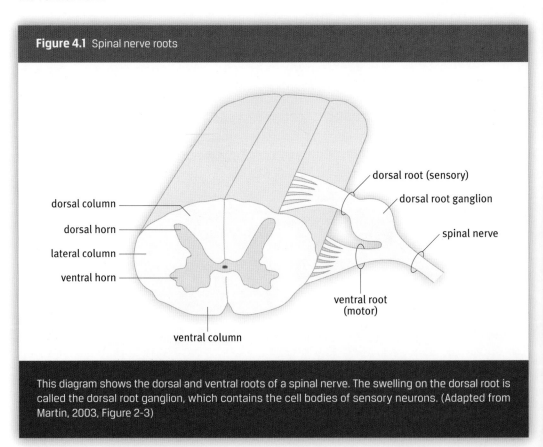

Figure 4.1 Spinal nerve roots

This diagram shows the dorsal and ventral roots of a spinal nerve. The swelling on the dorsal root is called the dorsal root ganglion, which contains the cell bodies of sensory neurons. (Adapted from Martin, 2003, Figure 2-3)

The midbrain and hindbrain begin with a similar arrangement (dorsally placed sensory structures and ventrally placed motor structures), but the addition of many other coordination systems makes it hard to distinguish this original pattern. Another difference in the region of the midbrain and hindbrain is that not all nerves have both motor and sensory components; some (like the nerves supplying the eye muscles) are specialized as purely motor nerves and others, like the nerves of balance and hearing, are purely sensory.

Somatic and visceral motor and sensory elements

In the spinal cord and hindbrain, the motor and sensory components can be further divided into somatic and visceral groups. The somatic groups supply voluntary (striated) muscles, and receive sensation from skin or vibration receptors. Visceral groups supply the internal organs, glands, and blood vessels, and receive sensation from internal organs and from taste receptors. The visceral motor elements in the spinal cord and hindbrain are often called autonomic, because of their automatic and involuntary functions. Autonomic elements can in turn be divided into sympathetic and parasympathetic groups. The sympathetic group is involved in emergency (fight or flight) responses, whereas the parasympathetic group controls the normal functions of the internal organs, such as salivation and digestion. The visceral motor and visceral sensory cells in the spinal cord and in the hindbrain are found approximately midway between the somatic motor and somatic sensory groups.

In the midbrain and hindbrain, there is a third group of motor elements–those that supply the muscles that were originally involved in the gill (branchial) arches. Although mammals do not have gills like fish, we have inherited the skeletal and muscular components of the gills of our swimming ancestors. The remnants of the first gill (branchial) arch form the jaw and the chewing muscles, the second arch forms the muscles of facial expression and a bone at the base of the tongue (the hyoid bone), and arches 3-6 form the muscles and bones of the pharynx and larynx.

Somatic
Somatic is a term referring to the skin, muscles and skeleton.

Visceral
Visceral is a term referring to internal organs, blood vessels, and glands.

Branchial structures
The term branchial refers to the gill arches of fish. The branchial arches are solid structures that separate the gill slits of fish. The remnants of the branchial arches are still present in mammalian embryos, forming parts of the head and neck. In mammals the first branchial arch forms the jawbone and chewing muscles.

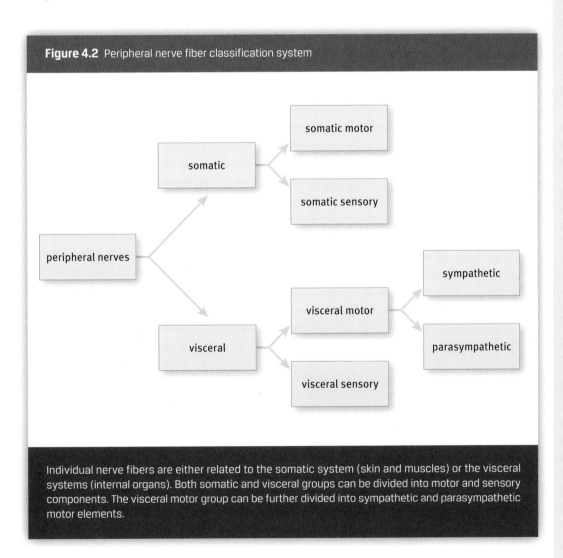

Figure 4.2 Peripheral nerve fiber classification system

Individual nerve fibers are either related to the somatic system (skin and muscles) or the visceral systems (internal organs). Both somatic and visceral groups can be divided into motor and sensory components. The visceral motor group can be further divided into sympathetic and parasympathetic motor elements.

Spinal nerves

The spinal cord is connected to the body by about thirty-two pairs of spinal nerves; the exact number varies from species to species. The nerves are named according the vertebral level at which they emerge, from rostral to caudal–cervical, thoracic, lumbar, sacral, and coccygeal. Each spinal nerve is formed by a sensory root and a motor root.

The pairs of spinal nerves arise from the spinal cord and leave the vertebral column through spaces between the vertebrae. Because the vertebral column grows faster than the spinal cord, the spinal cord in the adult human extends down to only the L1 or L2 vertebra. The result is that only the cervical segments of the spinal cord are approximately level with their corresponding vertebrae; lower down the spinal nerves run increasingly obliquely downwards to their exit points. At the end of the spinal cord (the level of the L2 vertebra) the remaining lumbar, sacral, and coccygeal nerves form a bundle called the cauda equina (the Latin name for a horse's tail).

Each of the dorsal and ventral roots of each spinal nerve is formed by 6-12 small rootlets. The fundamental difference between the ventral and dorsal roots was discovered by Magendie almost two hundred years ago. He showed that the dorsal roots contained sensory fibers, whereas the ventral roots contained motor fibers. When the dorsal and ventral roots join, they form a spinal nerve with both sensory and motor components.

The relationship between spinal nerves and the vertebrae is basically simple, but the naming system for the spinal nerves is not consistent. All the spinal nerves arising from thoracic, lumbar, and sacral segments of spinal cord are given the name of the vertebra immediately above their exit, but in the cervical region the rules differ because there are eight cervical spinal nerves but only seven cervical vertebrae. The problem has been solved in the following way; the C1 to C7 nerves are said to emerge above their respective vertebrae, and the C8 nerve exits between C7 and T1.

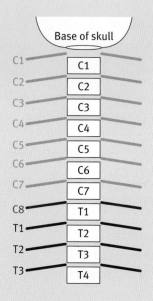

Figure 4.3
Naming of cervical spinal nerves

All thoracic, lumbar, sacral, and coccygeal spinal nerves are named for the vertebral body below which they emerge. The cervical nerves, on the other hand, are named for the vertebral body above which they emerge. The exception is the nerve that emerges between C7 and T1, which is called the C8 nerve.

Each dorsal root bears an ovoid swelling named the dorsal root (spinal) ganglion (Figure 4.1). Each dorsal root ganglion is home to thousands of sensory neurons, called pseudounipolar neurons. Each sensory neuron gives rise to a single axon, which then bifurcates, with one branch connecting to the periphery and the other connecting with the dorsal horn of the spinal cord.

In the trunk, each spinal nerve starts from the spinal cord and follows the line of the ribs, emerging from between the vertebrae and travelling in an arc until it reaches the ventral midline. Each spinal nerve supplies muscles and the overlying skin. The strip of skin supplied by a single spinal nerve is called a dermatome. The dermatomes of successive segments are represented as narrow bands of skin running almost horizontally along the trunk like a series of stripes. In the limbs there is a distortion in these segmental strips due to the elongation and rotation of limb buds during development, and the dermatomes are generally oriented longitudinally along the main axis of each limb.

The dermatomes were originally mapped by studying the loss of sensations in patients whose spinal nerve roots have been damaged, or by mapping the distribution of vesicles and blisters in cases of herpes zoster (shingles). In shingles, the virus is usually localized to a single dorsal root ganglion, and the characteristic rash is found in a single dermatome. Dermatome maps are useful in enabling the clinician to determine the location of pathology affecting the spinal cord or nerve roots.

Spinal nerves supplying the limbs

The limb buds in the fetus grow out from the trunk. Each limb bud is supplied by five spinal nerves. In the human these segments are C5, C6, C7, C8, and T1 for the upper limb and L2, L3, L4, L5, and S1 for the lower limb. The segments supplying each limb are very similar in all mammals, but there is some minor variation in the lower limb segments involved.

Spinal nerves supply the muscles, joints, and skin of each limb. The five spinal nerves of the upper limb (C5, C6, C7, C8, and T1) are rearranged in a plexus (a kind of interchange area) to form three main nerves, the median, ulnar, and radial nerves. The spinal nerves of the lower limb (L2, L3, L4, L5, and S1) similarly recombine into three main nerves; femoral, obturator, and sciatic nerves. The sciatic nerve is the largest nerve in the body, being over 1 cm thick in humans.

Cranial nerves

The brain is connected to the parts of the head and neck by twelve pairs of nerves, called the cranial nerves. Like the spinal nerves, the cranial nerves have motor and sensory components. However, because the head is very specialized compared with the rest of the body, the organization of the cranial nerves has some unusual features. First, some of the nerves are purely sensory–the nerves for olfaction (sense of smell), vision, hearing, and balance. Conversely, some of the nerves are purely motor– the nerves to the eye muscles and the tongue. The remaining nerves have both motor and sensory components.

Another complication in the cranial nerves is the fact that some of them were originally used to control the gill muscles of our distant fish ancestors. Mammals do not have gills, but the remnants of the gill muscles have been put to other uses–for chewing, facial expression, and swallowing. The gill motor neurons are called branchial motor neurons (branchial is just a technical adjective describing the gill of a fish). The branchial motor elements are difficult to classify, and some authors have attempted to classify them as either somatic motor or special visceral categories. To avoid this difficulty it is simpler to call them branchial motor. All the nerves that supply branchial arch remnants have both motor and sensory components. They are the 5th, 7th, 9th, 10th, and 11th cranial nerves.

Figure 4.4 Human dermatomes

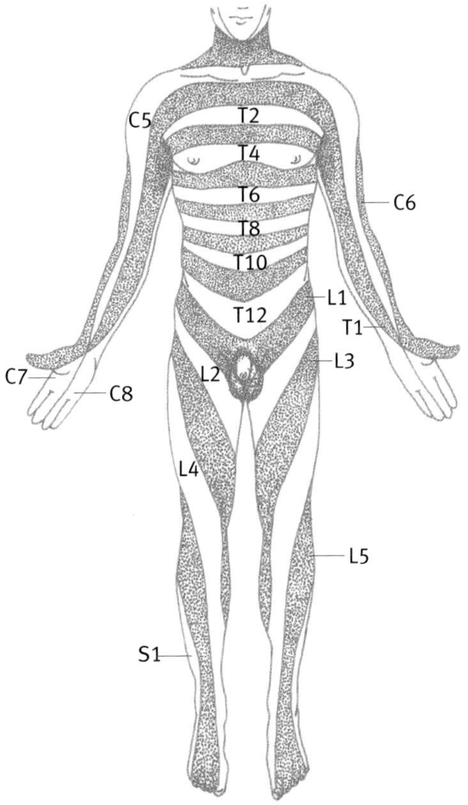

The area of skin supplied by a single spinal nerve is called a dermatome. In the trunk the dermatomes are approximately horizontal. In the limbs, the dermatomes have been stretched out by developing limb buds (Adapted from Cramer and Darby, 2005, p 345)

Mapping the dermatomes

The discovery and mapping of the dermatomes was based on the study of cases in which individual spinal nerves had been damaged or infected. Chicken pox virus can infect the neurons of a single dorsal root ganglion, and the painful vesicles appear in the region of a single dermatome. This condition is called shingles.

The olfactory and optic nerves

Despite their names, the olfactory and optic nerves are both tracts rather than nerves; they are formed from extensions of the developing forebrain. Part of the evidence for the special nature of these two nerves is that their axon myelin sheaths are made by oligodendroglia rather than by Schwann cells. Another characteristic is that the olfactory and optic nerves are both surrounded by three layers of meninges and cerebrospinal fluid, like other parts of the forebrain. Ordinary nerves are not enclosed this way. Despite this, we are forced by weight of tradition to keep the olfactory and optic extensions of the forebrain as cranial nerves 1 and 2.

The olfactory nerve (1)

The olfactory epithelium in the roof of the nose gives rise to fifteen rootlets, which enter the skull through a group of small holes in the ethmoid bone. As pointed out above, the olfactory nerve rootlets are part of the forebrain and each is surrounded by meninges and cerebrospinal fluid. This is of great significance in head injury that severs these rootlets, because cerebrospinal fluid leaks from the nose. The olfactory rootlets form the olfactory nerve, which then travels to the olfactory bulb. In rodents, the olfactory bulb is relatively huge, but in humans it is a pea-sized structure hidden under the frontal lobe of the cerebrum.

The optic nerve (2)

The optic nerve is formed by the axons of retinal ganglion cells. Because the optic nerve is really a tract, not a nerve, it is surrounded by the three layers of meninges and cerebrospinal fluid. The two optic nerves meet underneath the hypothalamus, just in front of the pituitary stalk, and many of the fibers cross to the opposite side. The crossing is called the optic chiasm and the fibers that leave the chiasm are called the optic tract. Some of the optic nerve fibers remain uncrossed; the uncrossed component is small (perhaps 10%) in rodents, but almost 50% in humans.

The cranial nerves of the midbrain and hindbrain

Cranial nerves 3-12 emerge from the midbrain and hindbrain as pairs of nerves. Unlike the first two cranial nerves, the remaining ten are true nerves, and are not covered by meninges.

Ganglion

A ganglion is a cluster of neuron cell bodies in the peripheral nervous system. The word ganglion simply means a swelling, but in the nervous system it always refers to a collection of neuron cell bodies.

Optic chiasm

The optic chiasm is the crossing of optic nerve fibers. It lies ventral to the rostral part of the hypothalamus. In humans, half the optic nerve fibers cross at the chiasm, but in rodents less than 10% cross the midline.

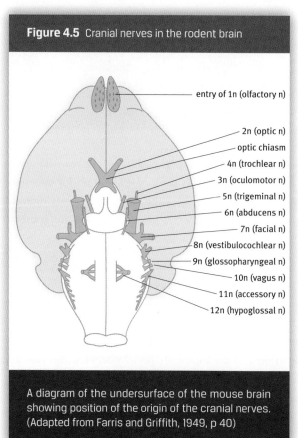

Figure 4.5 Cranial nerves in the rodent brain

- entry of 1n (olfactory n)
- 2n (optic n)
- optic chiasm
- 4n (trochlear n)
- 3n (oculomotor n)
- 5n (trigeminal n)
- 6n (abducens n)
- 7n (facial n)
- 8n (vestibulocochlear n)
- 9n (glossopharyngeal n)
- 10n (vagus n)
- 11n (accessory n)
- 12n (hypoglossal n)

A diagram of the undersurface of the mouse brain showing position of the origin of the cranial nerves. (Adapted from Farris and Griffith, 1949, p 40)

Purely motor cranial nerves–3, 4, 6, 12

Four cranial nerves are purely motor–the 3rd, 4th, 6th, and 12th.

- Cranial nerve 3 is called the oculomotor. It supplies four of the six muscles that move the eye. It also supplies the eyelid raising muscle and the muscles that focus the lens and constrict the pupil.
- Cranial nerve 4 is called the trochlear. It supplies the superior oblique muscle of the eye.
- Cranial nerve 6 is called the abducens. It supplies the lateral rectus muscle of the eye.
- Cranial nerve 12 is called the hypoglossal. It supplies the muscles of the tongue.

Purely sensory cranial nerves–1, 2, 8

- Cranial nerve 1 is the olfactory, and is formed from an outgrowth of the forebrain.
- Cranial nerve 2 is the optic, also an outgrowth of the forebrain, and includes the complex structure of the retina.
- Cranial nerve 8 combines the nerves of hearing (the cochlear nerve) and balance (the vestibular nerve). They are sensory nerves that are connected to receptors in the inner ear. Because they travel close together toward the hindbrain, they are given a collective name–the vestibulocochlear nerve.

Cranial nerves with both motor and sensory components–5, 7, 9, 10, 11

All the nerves that supply branchial arch remnants have both motor and sensory components.

- Cranial nerve 5 is called the trigeminal. It supplies the muscles of the jaw (the muscles of mastication) and the skin of the face.
- Cranial nerve 7 is called the facial. It supplies the muscles of facial expression and receives taste sensation from the front of the tongue. It also supplies the small glands of the head–the lacrimal, submandibular, and sublingual glands.
- Cranial nerve 9 is called the glossopharyngeal. It supplies one of the pharynx muscles and receives sensation from the back of the tongue and the surface of the pharynx. It also supplies the largest salivary gland, the parotid gland.
- Cranial nerve 10 is called the vagus. It supplies the muscles of the pharynx and the larynx and all of the organs of the thorax and abdomen, as far down as the brim of the pelvis. This includes the visceral motor and sensory supply of the lungs, heart, stomach, and almost all of the intestines.

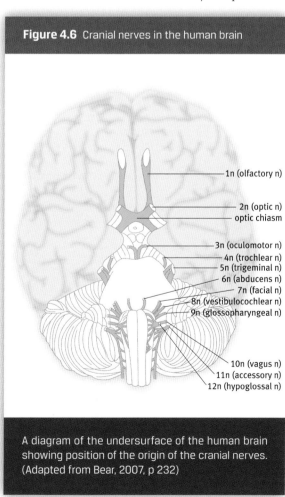

Figure 4.6 Cranial nerves in the human brain

A diagram of the undersurface of the human brain showing position of the origin of the cranial nerves. (Adapted from Bear, 2007, p 232)

The eye muscles

All vertebrates have six muscles to move each eye. The arrangement of the muscles allows the eye to move in any direction. The eye muscles are supplied by the three cranial nerves, the oculomotor, trochlear, and abducens. In humans, the two eyes always move in a coordinated way to look at the same point, but in animals where the eyes look sideward (such as frogs and lizards), the eyes can move independently. As well as the muscles that move the eyeball around, there are small internal eye muscles that control the size of the pupil and the focusing power of the lens. These internal eye muscles are supplied by the oculomotor nerve.

Table 4.1 The twelve cranial nerves

Number	Name	Functional components	Function
1	Olfactory	Special sensory	Sensation of smell.
2	Optic	Special sensory	Sensation of vision.
3	Oculomotor	Somatic motor	Controls four of the six eye muscles and the eyelid muscle.
		Visceral motor	Parasympathetic control of the lens and pupil.
4	Trochlear	Somatic motor	Controls the superior oblique muscle of the eye.
5	Trigeminal	Somatic sensory	Sensation from the skin of the face.
		Branchial motor	Controls the muscles of the jaw.
6	Abducens	Somatic motor	Controls the lateral rectus muscle of the eye.
7	Facial	Somatic sensory	Taste sensation from the front of the tongue.
		Branchial motor	Controls the muscles of the face.
8	Vestibulocochlear	Special sensory	Sensations of hearing and balance.
9	Glossopharyngeal	Branchial motor	Controls one of the muscles of the pharynx.
		Visceral motor	Parasympathetic control of the parotid salivary gland.
		Special sensory	Taste sensation from the back part of the tongue.
		Visceral sensory	Detection of blood pressure changes.
10	Vagus	Visceral motor	Parasympathetic control of internal organs of the thorax and abdomen.
		Visceral sensory	Sensation from the internal organs.
		Branchial motor	Controls the muscles of the pharynx and larynx.
11	Accessory (spinal part)	Somatic motor	Controls two large neck muscles.
12	Hypoglossal	Somatic motor	Controls the muscles of the tongue.

A cranial nerve mnemonic

Past generations of students have used the initials of the 12 cranial nerves to construct a memory aid (a mnemonic) in the form of a 12 word sentence. An example is 'Our Old Ostrich Tripped Twice After Fleeing Vicious Green Vipers And Hyenas'. Some other popular mnemonics of the past are more memorable than printable.

Cranial nerve 11 is called the accessory, because it assists the vagus in the supply of the pharynx. Cranial nerve 11 has a surprising additional component, which supplies motor nerves to two neck muscles. This component arises from the cervical spinal cord, travels up to the hindbrain to join the main accessory nerve briefly, and then leaves to run down to the two target muscles in the neck. This component is called the spinal accessory nerve.

Damage to the cranial nerves

In humans, the cranial nerves may be damaged by injury or inflammation, producing characteristic problems. This section is a summary of some important examples of cranial nerve damage.

Damage to the oculomotor, trochlear, and abducens (cranial nerves 3, 4, and 6)

If one of these nerves is damaged, the movement of the two eyes cannot be coordinated. The result is that the affected eye does not look directly at the object of interest. This condition is called strabismus or squint. Because the oculomotor (3rd) cranial nerve also supplies the eyelid, pupil and lens, damage to this nerve also results in a drooping eyelid, a dilated pupil, and an inability to focus the lens for close-up objects.

Damage to the trigeminal nerve

The trigeminal nerve (5) is sometimes infected with the herpes zoster (shingles) virus. Painful vesicles appear in the face, usually in the area of the forehead and upper eyelid. This may clear up, but sometimes a severe and debilitating nerve pain persists for many months. The condition can be treated successfully with antiviral drugs if the treatment is started very early.

Damage to the facial nerve

The facial nerve (7) is commonly damaged by a viral infection. The resulting paralysis is called Bell's palsy. The sufferers often report loss of taste sensation in the front two-thirds of the tongue. The paralysis usually clears up, but in about a third of cases, the nerve does not recover.

Damage to the vestibulocochlear nerve

Damage to the vestibulocochlear nerve (8) can be caused by infections, drugs, tumors, and an inflammatory condition called Menière's disease. In severe cases the result is permanent deafness and loss of balance and vertigo. In less severe cases there may be temporary tinnitus (ringing or buzzing sensations in the ear) and vertigo.

Damage to the hypoglossal nerve

When the hypoglossal nerve (12) is damaged, the protruded tongue deviates to the side of the damaged nerve.

The autonomic nervous system

The autonomic nervous system includes the motor neurons that control the physiology of the viscera, sweat glands and blood vessels. A characteristic feature of the autonomic nervous system is that its effector (motor) neurons are located outside the central nervous system. They are found in a string of swellings (ganglia) lying at the side of the vertebral column. Sensory fibers coming from visceral organs accompany the axons of the autonomic motor neurons.

Figure 4.7
Oculomotor nerve palsy

When the oculomotor nerve is damaged the eye is pulled laterally and downwards by the two remaining eye muscles (supplied by the fourth and sixth cranial nerves). The oculomotor nerve also supplies the muscle of the upper eyelid and the constrictor of the pupil, so that oculomotor nerve damage results in a drooping eyelid and a dilated pupil.

Figure 4.8
Facial nerve palsy

When the facial nerve is damaged the muscles of facial expression are paralyzed. The most obvious sign is the drooping of the corner of the mouth on the affected side.

Palsy

Palsy is an old word for paralysis. It is still commonly used by neurologists, who use facial palsy as the name for facial nerve paralysis. Parkinson's disease was originally called the 'shaking palsy', because of the combination of tremor and movement restriction.

The autonomic nervous system consists of two major functional divisions—parasympathetic and sympathetic—each with characteristic anatomy and functions. Both branches of the autonomic nervous system are regulated by the hypothalamus, and to a lesser extent by hindbrain nuclei such as the solitary nucleus. However, many autonomic reflexes operate independent of the central nervous system. This independence is reflected in the name autonomic, which means self-directing.

Sympathetic nervous system

The sympathetic nervous system was so-named because its actions are often grouped together in 'sympathy' with the needs of the body. For example, in stressful situations the heart rate increases, skin blood vessels constrict, and the gut slows down in one coordinated reaction. This occurs in part because of the anatomy of the sympathetic outflow. The sympathetic nervous system is structured in such a way that the sympathetic motor neurons each have an influence over a wide area. When their axons leave the spinal cord, they enter clusters of neurons called the sympathetic chain or paravertebral ganglia. Most of the sympathetic motor axons terminate in these ganglia, exciting large networks of ganglionic motor neurons. The ganglionic motor neurons then send post-ganglionic axons to the viscera, or via the spinal nerves to sweat glands and blood vessels in

The adrenal medulla

The medulla (central part) of the adrenal gland acts like a huge sympathetic ganglion, producing very large amounts of adrenaline for release directly into the bloodstream. The secretions of the adrenal medulla help to amplify, expand, and sustain the response of the sympathetic nervous system to stress, by maintaining high circulating levels of adrenaline. It is the largest neuroendocrine effector system.

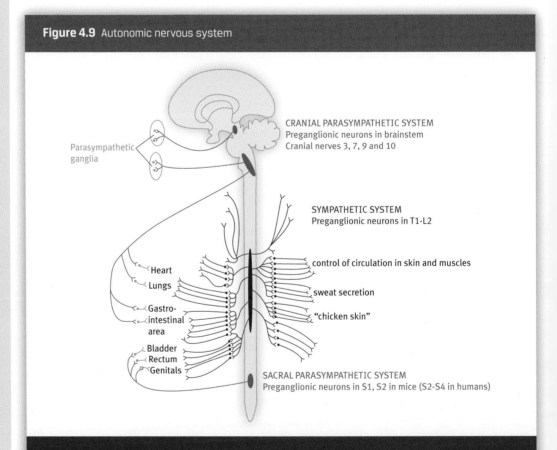

Figure 4.9 Autonomic nervous system

This diagram shows the origin and course of the sympathetic preganglionic neurons (colored red) and parasympathetic preganglionic neurons (colored blue). The postganglionic neurons are colored pink and pale blue. The parasympathetic preganglionic neurons are located in the midbrain and hindbrain, and in the sacral spinal cord. The sympathetic preganglionic neurons are located in the thoracic and upper lumbar spinal cord. (Adapted from Brodal, 2004, p 372)

the periphery, where they control sweating and skin blood supply during stress and exertion. Some of the other sympathetic motor axons pass through the paravertebral ganglia and emerge to activate motor neurons in ganglia closer to the target tissue (celiac, superior mesenteric and inferior mesenteric ganglia); others directly stimulate the adrenal medulla to produce adrenaline, which circulates in the bloodstream and causes whole-body sympathetic effects.

Parasympathetic nervous system

The parasympathetic nervous system facilitates the normal day-to-day functions of the internal organs, such as digestion and bladder emptying. The parasympathetic preganglionic motor neurons are found in two groups, one in the brainstem and one in the sacral spinal cord. The brainstem component is distributed through cranial nerves 3, 7, 9, and 10. The parasympathetic motor neurons send their axons to effector neurons. Unlike the sympathetic system, parasympathetic effector neurons are generally not clustered in ganglia. Instead they are either on the surface of visceral organs or embedded in the walls of organs. Despite the absence of visible parasympathetic ganglia in many cases, we will still refer to the axons of parasympathetic effector cells as postganglionic axons in order to help with comparison with the sympathetic system. The postganglionic fibers that arise from the parasympathetic effector neurons are very short, and they are restricted to the visceral organs. Unlike the sympathetic nerves, parasympathetic fibers do not travel in spinal nerves to innervate other peripheral tissues.

Bladder control

The centers for bladder control are located in the hindbrain near the locus coeruleus. One center (the Barrington nucleus) commands the bladder to empty, and a nearby continence center can stop the flow of urine. The bladder-emptying commands are sent to the sacral parasympathetic system. The continence center commands are sent to special motor neurons in the sacral spinal cord ventral horn, which cause the sphincter of the urethra to contract and stop the flow of urine.

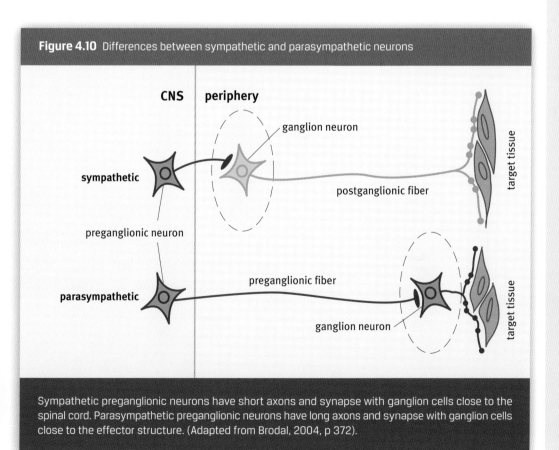

Figure 4.10 Differences between sympathetic and parasympathetic neurons

Sympathetic preganglionic neurons have short axons and synapse with ganglion cells close to the spinal cord. Parasympathetic preganglionic neurons have long axons and synapse with ganglion cells close to the effector structure. (Adapted from Brodal, 2004, p 372).

Enteric nervous system

The enteric nervous system (ENS) is a web of sensory neurons, motor neurons, and interneurons embedded in the wall of the gastrointesinal system, stretching from the lower third of the esophagus right through to the rectum. The neurons of the ENS are arranged in two layers, the submucosal and myenteric plexuses of the gut wall. It has been estimated that the ENS actually contains more neurons than the whole of the spinal cord. The ENS processes a range of sensations, such as the nature of gut contents and gut distension, and integrates this information with input from the autonomic nervous system. In this way the ENS can guide and optimise the muscular and secretory activity of the gastrointestinal tract. Many of the ENS effector neurons are also innervated by parasympathetic motor neurons, so they act as effector neurons of the parasympathetic nervous system. For this reason the ENS is regarded as an integral part of the parasympathetic nervous system, but its specialized sensory neurons and independent processing make it rather more complex than a simple parasympathetic ganglion. The ENS displays sophisticated coordination and exhibits plasticity and learning in response to changing dietary habits or disruptions to the gut.

Chapter 5
Command and control—the motor systems

5 Command and control – the motor systems

In this chapter:

Command and control of skeletal muscles	56
Areas of the motor cortex	57
Survival skills: the hypothalamus	59
Brainstem and spinal cord modules for control of organized movement	60
Descending control pathways other than the corticospinal tract	61
The role of the cerebellum in motor control	61
The roles of the striatum and pallidum in motor control	64
The final common pathway for all motor systems–the motor neuron	66
Command and control of the visceral systems–the autonomic nervous system	68
Command and control of the neuroendocrine system	72

A major role of the nervous system is to initiate actions in response to challenges in the environment. These responses include movement of the body by muscle contraction, internal adjustments to heart rate and breathing, and secretion of hormones in response to stress and other bodily requirements.

The systems that get things done can be called the effector systems. They give commands, and they control the activities they initiate. The effector systems can be divided into those that command and control skeletal muscles, those that command and control internal organs, and those that command and control endocrine glands.

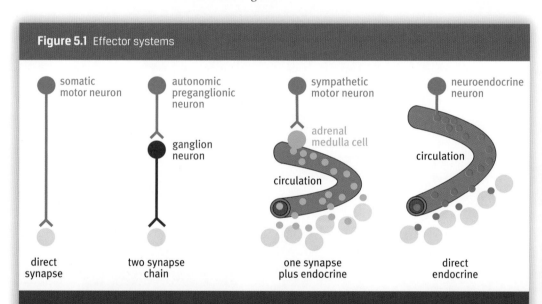

Figure 5.1 Effector systems

This diagram shows four different kinds of effector systems. The somatic motor neuron makes a direct synapse with the effector cell–in this case a skeletal muscle cell. The sympathetic and parasympathetic motor neurons synapse with a ganglion cell, which then synapses with the effector cell (such as a smooth muscle cell). Some sympathetic motor neurons synapse with modified ganglion cells in the adrenal medulla. These modified ganglion cells have no axon and the transmitter substance (adrenaline) is secreted directly into the blood stream. Neuroendocrine motor neurons in the hypothalamus secrete transmitter molecules directly into the blood stream. This is a direct endocrine action. (Adapted from Swanson, 2003, p 99)

Command and control of skeletal muscles

The skeletal muscles are under conscious control most of the time. For this reason, they are often called the voluntary muscles–those that contract when we want them to. However, in some cases the contraction of these muscles is reflexive, such as when we burn our fingers on a hot stove. Reflexes operate more quickly than the normal system of conscious control.

In humans, the most important system for control of activity of the skeletal muscles starts in the motor area of the cerebral cortex, which in humans lies just in front of the central sulcus. Stimulation of this area produces muscle contraction on the opposite side of the body. These experiments have shown that there is a somatotopic organization in the motor area of the cortex. This somatotopic map represents the position of muscles, with the same basic arrangement as the parts of the body (foot, leg, thigh in sequence). The map is draped across the surface of the precentral gyrus, with the foot and leg on the medial surface of the hemisphere, and the thigh, trunk, arm and face running across the supero-lateral surface of the precentral gyrus. The whole motor map is upside down in relation to the body.

A large fiber tract, called the corticospinal tract, connects the motor cortex with the spinal cord. The corticospinal tract arises from large cells in the motor cortex (and a large number of other areas) and sequentially passes through the internal capsule, the cerebral peduncle, the pons, and the medullary pyramid to reach the caudal end of the hindbrain. At this point, most of the corticospinal fibers cross to the opposite side in the pyramidal decussation. They make connection with motor neurons on the opposite side of the body from their cortical origin.

Figure 5.2 Somatotopic organization of the precentral gyrus

The human motor cortex is located in the precentral gyrus. The parts of the body are represented upside down, with the lower limb at the top and the mouth at the bottom of the gyrus. The orderly arrangement of the parts of the body represented on the precentral gyrus is called a somatotopic map. (Adapted from Bear et al, 2007, p 402)

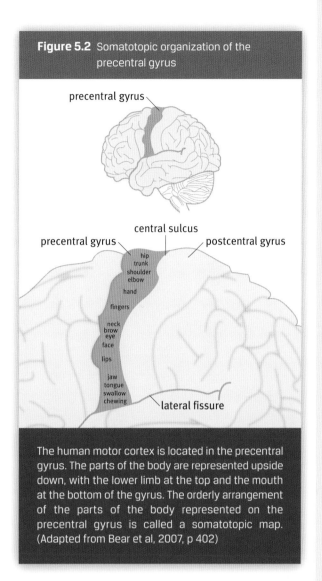

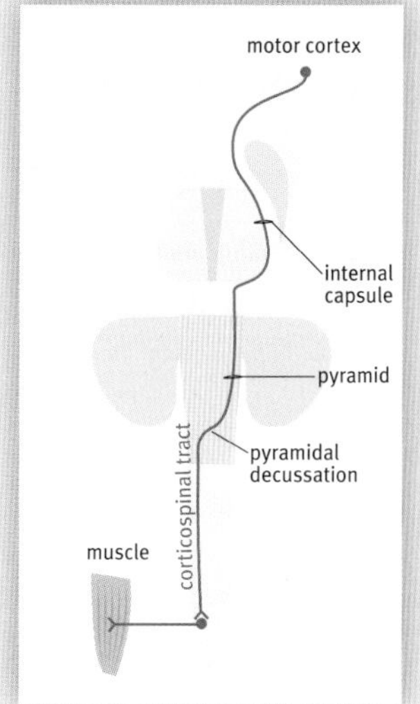

Figure 5.3
Corticospinal tract

The corticospinal tract arises chiefly from large neurons in the motor cortex. The axons travel through the internal capsule, cerebral peduncle, and pons to reach the pyramid of the hindbrain. At the junction of hindbrain and spinal cord the corticospinal fibers cross the midline and travel down the spinal cord. They synapse with motor neurons on the side opposite to their cortical origin. (Flat map adapted from Swanson, 2004)

Changing names: corticospinal tract or pyramidal tract?

Because the corticospinal fiber bundles form the medullary pyramids in mammals, they have traditionally been called the pyramidal tracts. This term has potential to confuse: by coincidence, the cells of origin are pyramid-shaped neurons in the cerebral cortex. Another reason to avoid the term 'pyramidal tract' is the unfortunate and now discredited 'pyramidal/extrapyramidal' concept, which attempted to explain the organization of different parts of the motor system. For these reasons we prefer the term corticospinal tract.

Areas of the motor cortex

The primary motor cortex is the source of most of the corticospinal axons, but its activity is strongly influenced by the pallidum, striatum, cerebellum and many other cortical regions, including the somatosensory area of the cortex. In some ways the activity of the primary motor cortex is a committee decision, reached by combining many influences, and enabled by the striatum and pallidum. The secondary motor cortex in humans is called the supplementary motor area. It is immediately in front of the lower limb area of the primary motor cortex, and so is located mostly on the medial surface of the cerebrum. Many movements are planned and

Striated and smooth muscle

The term 'striated' muscle is commonly used as a synonym for 'voluntary muscle.' The reason is that the fibers of the muscles of the limbs and body wall, which are under voluntary control, appear striped under the microscope. The involuntary muscles of the intestines and bladder do not have striations and are called 'smooth' muscle fibers. However, the heart muscle fibers are striated even though they are not under voluntary control.

sequenced in the secondary motor cortex. These libraries manage planning and sequencing of complex movements. An example is Broca's region, which contains movement patterns for making speech sounds and writing. In 90% of people, Broca's region is in the left frontal lobe, and these people are almost all right-handed. In left-handed people, the Broca's area is found in either the right or left frontal lobe. The remainder of the frontal lobe is called the prefrontal cortex; the prefrontal cortex is related to the generation of abstract thought and complex social behaviors.

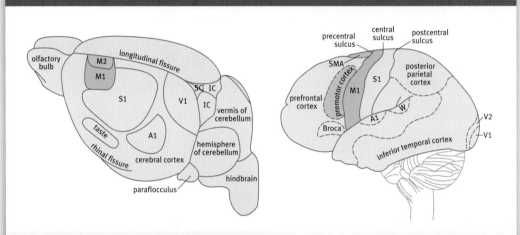

Figure 5.4 Areas of the motor cortex

This picture shows the brain areas involved in motor control in the rat and human. In the rat the primary motor cortex (M1) is rostral to S1, and the secondary motor cortex (M2) is medial to M1. In the human, M1 is once again rostral to S1, but M2 is called the supplementary motor area (SMA). The SMA has been pushed into the medial side of the hemisphere, and only a small strip of SMA can be seen rostral to M1 in this picture. In humans the frontal lobe contains a number of additional motor areas, the premotor area and Broca's area.

The human frontal lobe

The frontal lobe of the cerebrum is greatly enlarged in humans. Like other primates, the human frontal lobe contains the motor areas. However, most of the human frontal lobe has functions associated with complex behaviors, related to social functioning and individual personality. The frontal lobe of the dominant hemisphere contains the speech area of Broca.

What happens when the corticospinal tract is damaged?

When the corticospinal tract on one side is cut at the level of the hindbrain in a monkey, the animal can move around in a normal fashion, and can use the hands for climbing. However, the monkey finds it almost impossible to pick up objects with the affected hand (the hand on the opposite side to the damage). Some weeks after the operation, the ability of the hand to grasp objects may improve, but the movements are clumsy and involve flexion of all the fingers, rather than one at a time. The ability to move the fingers independent of each other and opposition of the thumb is permanently lost. This shows that fine movements of the fingers and thumb are exclusively related to the corticospinal tract.

Humans who experience damage to the motor cortex or the corticospinal tract are unable to perform tasks that require precise movements of individual fingers, such as writing, sewing, tying knots, and picking up small objects. Less precise movements, and those that involve larger muscle groups, are less severely affected. When the corticospinal tract is damaged by large forebrain lesions in humans, the loss of fine movement is accompanied by hypertonia (stiffness or spasticity) and hyper-reflexia. These deficits do not appear in monkeys with lesions restricted to the corticospinal tract, so they are probably due to damage to adjacent forebrain structures, such as the striatum, in humans.

The role of non-cortical motor centers

We plan movements conceptually, thinking in sequences and outcomes rather than joint angles and individual muscles. The brain translates these high-level plans into the initiation of contraction of dozens of muscles needed to carry out the required movements. We do not have to think of the individual muscles; we only have to think of what we want to happen. For instance, scratching the back of the neck involves coordinated movements of the wrist, elbow, shoulder and finger joints in a complicated sequence, even though the idea of the movement is very simple. The motor systems in the brain and spinal cord are arranged in such a way that movement 'requests' are translated into complex actions, without us having to deal with the detailed coordination of dozens of individual muscles.

In humans, the corticospinal tract is vitally important for delicate movements of the fingers, but most of our routine activities are handled by functional systems in the brainstem and spinal cord, which are called modules. These modules are wired during brain development to produce semi-automatic movements like walking and chewing. This makes it easier for the motor cortex to get a result without the need to connect to individual motor neurons. During vertebrate evolution, complex behaviors and movement sets became organized in a hierarchy of modules. At the top of the hierarchy are eating and drinking behaviors, defensive behaviors, reproductive behaviors, and the movement patterns to explore the immediate environment. These high-level behaviors are vital for the survival of the individual and the species.

Survival skills: the hypothalamus

This important group of survival behaviors is primarily controlled by the hypothalamus. The hypothalamic control centers are clustered in two main groups. The first group includes systems for defense, reproduction, and ingestion (feeding and drinking). The nuclei that control these behavior sets are located in the medial half of the hypothalamus, including the large ventromedial nucleus (VMH). Each specific behavior set, such as eating and drinking, is related to a specific set of hypothalamic nuclei.

The second group of survival behavior organizers is located in the mammillary body in the caudal part of the hypothalamus, and also in adjacent parts of the midbrain (the substantia

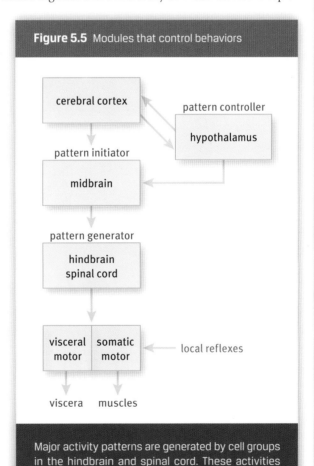

Figure 5.5 Modules that control behaviors

Major activity patterns are generated by cell groups in the hindbrain and spinal cord. These activities include locomotion, breathing, and swallowing. The activity patterns often have visceral and somatic components. The pattern generators are in turn controlled by higher-level centers in the cerebral cortex, hypothalamus, and midbrain. (Adapted from Swanson, 2007, p 111)

The hypothalamus

The hypothalamus is one of the four major subdivisions of the embryonic forebrain. The other parts of the forebrain are the telencephalon, the diencephalon, and the eye. The hypothalamus used to be considered part of the diencephalon, but gene expression studies reveal that it is a separate part of the forebrain, rostral to the diencephalon.

nigra and the ventral tegmental area). This second group initiates exploratory and foraging behaviors, such as turning the head and eyes toward an object of interest, and walking or running to explore the surroundings.

Brainstem and spinal cord modules for control of organized movement

For example, feeding may involve turning the head to look around, running to a food source, reaching and grasping, and licking and chewing. Each of these activities is sequenced by a particular part of the midbrain, hindbrain, or spinal cord. The hypothalamus activates a number of these movement modules in a coordinated program to achieve the desired behavior. The brainstem offers modules for orienting movements of the head and neck, licking, chewing, facial expression, vocalization, and control of breathing. The movement modules in the spinal cord include those that control posture, locomotion, reaching, and grasping. In the spinal cord, function-specific groups of interneurons in the gray matter activate primary motor neurons with precise, but adjustable, sequencing and timing, producing simpler movements involving groups of muscles and motor neurons. A simple example of modular control is withdrawal of the lower limb from a sharp pain in the foot. The movement happens before the pain is even registered in our cerebral cortex. The main response is to pull the foot away from the source of the pain, while

Modules that create movement patterns

Humans are very aware of the movements that are voluntarily directed, such as writing with a pen or kicking a football. However, most of our movement patterns are only semi-voluntary. Walking is a good example. We voluntarily commence walking, but after that the spinal cord takes over. The spinal cord contains a program or module (in the form of a nerve network) that controls all the muscles required for walking, and we can continue to walk without having to think about the individual movements involved.

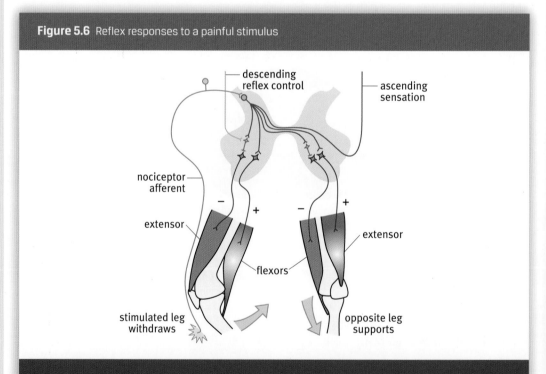

Figure 5.6 Reflex responses to a painful stimulus

A painful stimulus (left) detected by a nociceptor afferent (see Chapter 6) has been relayed an 'interneuron' in the spinal cord. This interneuron influences a number of motor neurons (red). On the left it activates the flexor withdrawal reflex to withdraw the leg, and inhibits an extensor motor neuron via an inhibitory interneuron (blue). On the right side, it activates the crossed extensor reflex to adjust posture. Descending pathways (orange line on the left) also influence the reflex circuit. The painful sensation is transmitted to the brain via an ascending tract (right). Note that the 'interneuron' is a diagrammatic simplification of a network involving many interneurons. (Adapted from Kandel and Schwarz, 2001).

inhibiting the corresponding extensors. The problem with this response is that it will also cause a fall, unless adjustments are made. The module takes care of this by contracting hip extensors, hip abductors, and knee extensors of the opposite leg, in order to stabilize the body. In this way, ten or more muscles in each leg are either contracted or made to relax, in order to achieve the whole pattern of movement needed to terminate a painful stimulus. The spinal cord has evolved many of these complex sub-systems during vertebrate evolution. The advantage of having these functional modules is that higher centers, such as the motor cortex or brainstem pattern generators, are only required to stimulate a few modular groups in order to initiate a whole pattern of movement.

Descending control pathways other than the corticospinal tract

As well as sending fibers to the spinal cord in the corticospinal tract, the motor cortex sends out corticopontine and corticobulbar connections to the cerebellum and brainstem motor centers, activating other parts of the somatic motor system. A number of important centers in the brainstem give rise to descending tracts that end in the interneuron pools of the intermediate spinal gray matter. These tracts include the tectospinal, rubrospinal, reticulospinal, and vestibulospinal tracts, each of which controls different aspects of movement.

Noradrenaline neurons of the locus coeruleus and serotonin neurons of the raphe also send pathways to the spinal cord. They are not involved in initiation of specific movements and their exact role is not clear. They probably have a general influence over the level of neuronal activity in the spinal cord.

> **Changing names: The extrapyramidal system**
>
> The concept of an extrapyramidal system became popular among neurologists who were trying to distinguish the motor problems arising from damage to the pyramidal (corticospinal) tract from other motor syndromes. They contrasted the direct (monosynaptic) projection of the cortex to motoneurons (the pyramidal system) with what they believed to be an alternative multisynaptic pathway from cortex to motoneurons. They hypothesized that this multisynaptic pathway went from the cortex to striatum, pallidum, various midbrain and hindbrain centers, and finally to the spinal cord. The problem with this concept is that the hypothesized chain of connections does not exist. In 1962, Nauta and Mehler showed that the globus pallidus projects almost entirely to the thalamus rather than to midbrain and hindbrain motor centers. Despite this, the 'extrapyramidal' concept lives on in textbooks and in lectures given by neurologists.

The role of the cerebellum in motor control

The cerebellum is responsible for coordination of complex movements. It does this by comparing the motor commands with the actual body movements that result. It can then adjust motor control to improve accuracy. This monitoring and feedback happens in real time, and is most important at the end stages of a movement. The cerebellum receives inputs from the vestibular system so that it can monitor the position of the body in space. The cerebellum also uses this information to monitor posture and balance, and makes major postural adjustments if we are in danger of losing our balance.

In primates, the cerebellum is vital for coordinating the movements of the hands and eyes. For monkeys, life in the trees depends entirely on the ability to plan and execute acrobatic leaps,

---

The red nucleus and the rubrospinal tract

The rubrospinal tract arises from the red nucleus of the midbrain. All fish, reptiles, birds, and mammals have a red nucleus, and we presume that they all have a rubrospinal tract. The red nucleus is an important motor center, controlling movements of the limbs. The red color of the red nucleus is due to deposits of iron.

The raphe nuclei

A number of neuron groups in the midline of the hindbrain manufacture serotonin (5HT). The axons of these serotonin neurons project widely over the brain and spinal cord. Serotonin has a number of different actions. It is involved in the regulation of mood, the control of pain, and in the regulation of the sleep-wake cycle. Many popular antidepressant drugs are designed to enhance the action of serotonin.

---

involving the cerebellum, eyes, and hands. Humans have the most complete version of binocular vision and have opposable thumbs. The human cerebellum has evolved to capitalize on these capacities through the precise coordination of skilled finger movements. The fine coordination of finger movement under the control of complete binocular vision has given humans access to tool manufacture and communication.

The cerebellum develops mainly after birth. In humans, the cerebellum is relatively tiny at the time of birth, but it grows rapidly after birth. During the years from one to six, the cerebellum systematically codes the movement sequences that make up the basic repertoire of running, climbing, jumping, and so on.

The structure of the cerebellar cortex is extraordinarily uniform. The same three layers are seen in all parts of the cerebellum and their relative thickness does not vary. The layer of large Purkinje cells separates the deep granule cell layer from the superficial molecular layer. Deep to the cortex of the cerebellum is a core of white matter. Embedded in this white matter are three pairs of cerebellar nuclei (lateral, interposed, and medial), which are the final output centers for the cerebellum.

The Purkinje cells are the output cells of the cerebellar cortex. Their axons project to the cerebellar nuclei, which are buried in the cerebellar white matter. The most lateral cerebellar nucleus, the dentate nucleus, is much larger in primates than in other mammals, and is huge in humans. The cells of the cerebellar nuclei mainly send their axons to the thalamus (to nuclei that then project to the cortex), but also to the brainstem and spinal cord.

Most of the input to the cerebellum either goes to the granule cells; a smaller pathway goes to the huge dendritic trees of the Purkinje cells. The input to the granule cells comes from the pontine nuclei and other large precerebellar nuclei of the brainstem. The fibers from these nuclei that terminate in the granule cell layer are called mossy fibers. The granule cells send their axons to Purkinje cells. The granule cells are tiny, but very tightly packed, and comprise about 70% of all the neurons in the central nervous system.

One large hindbrain nucleus is an exception in the way it projects to the cerebellum. This is the inferior olivary nucleus, whose axons do not contact granule cells. Instead they form climbing fibers, which make direct contact with the dendritic trees of the Purkinje cells. There is evidence that the inferior olivary cells control the time code of the cerebellum, thus ensuring that every learned movement is played back in exactly the right timing sequence. The inferior olive cells fire at about ten action potentials per second.

Microelectrode recordings in the rat suggest that every folium in the cerebellum codes a different movement sequence. For example, some are concerned with taking food with the fingers and transferring it to the mouth, and others are involved in running behavior. Given the tens of thousands of folia in the human cerebellum, it is possible that the cerebellum keeps learning new movement sequences into adult life. Perhaps there is even a backhand folium in keen tennis players.

The cerebellum is connected to the vestibular (balance) system, the spinal cord, the midbrain, and the cerebrum. The largest connection by far is that with the motor cortex. A huge pathway arises from the cerebral cortex, synapses in the pons, and enters the cerebellum through the middle cerebellar peduncle. This seems to be a way for the motor cortex to ask for help from the cerebellum when a complex movement is required. The answer comes back to the motor cortex via the superior cerebellar peduncle. The fibers of the superior cerebellar peduncle synapse in the thalamus and the thalamocortical fibers travel to the motor cortex. The relationship between cell numbers in the cerebral cortex and the cerebellum is very consistent across the mammalian groups.

Purkinje cells

The Purkinje cells form a single-cell layer separating the molecular and granular layers of the cerebellum. Each Purkinje cell has a very large dendritic tree, receiving about a half a million synapses. The Purkinje cell dendritic tree is fanned out in a two-dimensional structure, like a paper cut-out tree. The tree lies at right-angles to the long axis of the cerebellar folium in which it is situated.

Figure 5.7 Cerebellar cortex

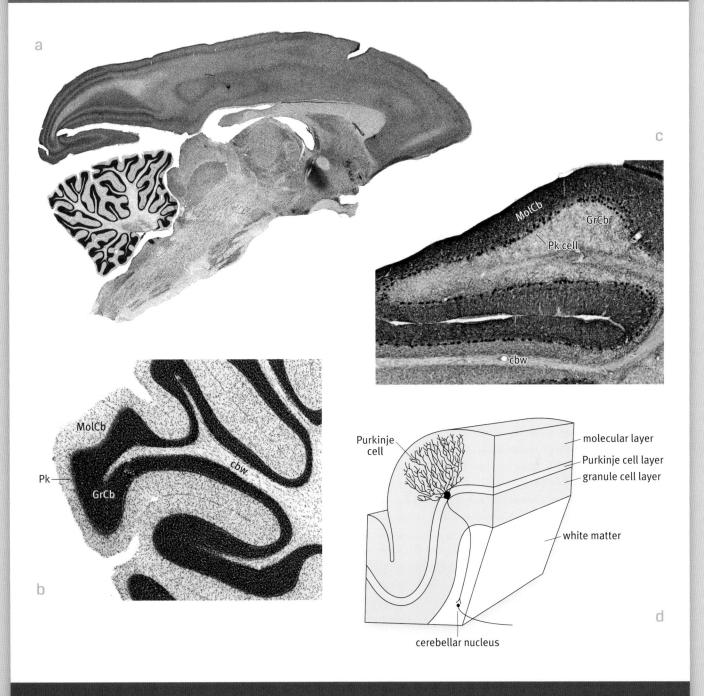

(a) A sagittal section through the brain of a monkey showing the intensely folded cerebellar cortex lying between the cerebral cortex (above) and the stem of the hindbrain (below).

(b) The cerebellar cortex consists of a pale outer layer (molecular layer) and a very dense inner layer (granular cell layer). Deep to the granular cell layer is the white matter of the cerebellum. The Purkinje cell layer is located between the molecular and granular layers but is not easy to define in this photograph.

(c) In this calbindin stained section the Purkinje cell bodies appear as black dots. The molecular layer is stained dark brown and the granular cell layer is lightly stained. The cerebellar white matter is unstained.

(d) A diagram showing the position of a Purkinje cell in relation to the layers of the cerebellar cortex. The Purkinje cell dendrites spread out in the molecular layer; the cell body is in the Purkinje cell layer and the axon travels through the granule cell layer to reach neurons in a cerebellar nucleus. (Adapted from Bear et al, 2007, p 773)

Although cerebral size varies a great deal from one family to another, the relationship between the number of cerebral cortical cells and the number of cerebellar cells is the same–about 1:4.

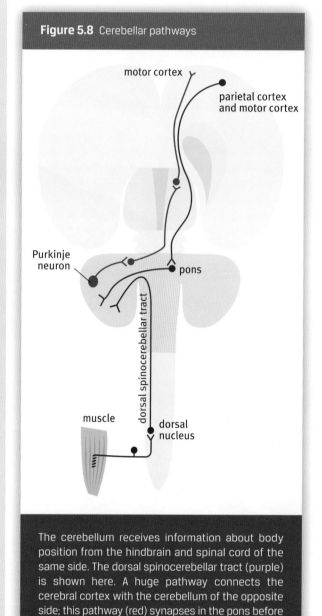

Figure 5.8 Cerebellar pathways

The cerebellum receives information about body position from the hindbrain and spinal cord of the same side. The dorsal spinocerebellar tract (purple) is shown here. A huge pathway connects the cerebral cortex with the cerebellum of the opposite side; this pathway (red) synapses in the pons before crossing. The Purkinje cells of the cerebellum can respond to the cerebral cortex via a pathway (green) which synapses in the cerebellar nuclei and the thalamus. (Flat map adapted from Swanson, 2004)

Striatum and pallidum

The striatum and pallidum are large masses of gray matter buried deep in the cerebrum. Developmentally, they are derivatives of the subpallium of the forebrain. They are both involved in motor control. The main parts of the striatum are the caudate, putamen, and accumbens nucleus. The main part of the pallidum is the globus pallidus.

The main way the cerebellum tracks movement is from sensations from the skin. It used to be thought that the most important information on movement would come from joint receptors and muscle stretch receptors, but it now appears that most of the input comes from receptors that detect sensations of stretching, contact, or folding in the skin. These sensations provide very fine detail on movement of the limbs and other parts.

The roles of the striatum and pallidum in motor control

The striatum and the pallidum are large masses of gray matter enclosed by the cerebral cortex in all mammals. These same groups have a long evolutionary history and can be clearly identified in reptiles and in birds. The striatum and pallidum are part of a chain of connections which regulate the form and style of movements. Pyramid-shaped cells in layer 5 of the cortex project to cells in the striatum, which in turn project to the pallidum. The pallidum projects to the midbrain and to the thalamus. The thalamic nucleus involved then projects back to the cerebral cortex, altering movements in the planning stage or while the movement is in progress. The substantia nigra is intimately connected with the striatum and also influences movements through this system.

Changing names: the basal ganglia

The term basal ganglia originally referred to the striatum, pallidum, amygdala, and thalamus. These days it is most commonly used to refer to the striatum and pallidum only, but the usage is inconsistent. For this reason, it is a term to use with care. In most cases it is better to refer directly to the striatum or pallidum or both, and avoid the use of the term basal ganglia.

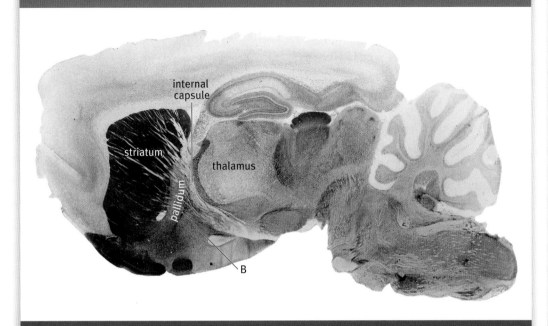

Figure 5.9 Striatum and pallidum in the rat

In this sagittal section of the rat brain the striatum (caudate and putamen) is stained dark brown with AChE (acetylcholinesterase stain). The pallidum is represented by a pale staining area (globus pallidus) between the striatum and the internal capsule. Along the caudal border of the globus pallidus is a line of very darkly staining cells called the basal nucleus. Note that the striatum has a number of foot-like extensions that reach the surface of the brain. The position of the basal nucleus (B) is shown.

The role of this prominent set of connections seems to lie in the production of predetermined movement patterns. Many species are known to have complex patterns of movement associated with mating and with displays of aggression. It seems possible that the striatal/pallidal complex is the main center for initiation of these genetically determined movement patterns. Other expressions of emotional states, such as smiling, also originate in the striatum and pallidum, and we are very sensitive to the emotional states these movements represent. 'Faking' a smile with conscious movements uses cortical motor control instead, and produces a result that looks false to others in most cases.

An interesting feature of the cells of the striatum and pallidum is that they influence the cortex by inhibiting thalamic neurons using the inhibitory neurotransmitter GABA (gamma aminobutyric acid). This makes them very different from the cortical neurons that connect to the striatum; these cortical neurons release the excitatory transmitter substance glutamate. This arrangement is the same in all mammals that have been studied.

We do not have a clear idea of the precise role of the striatum and pallidum in humans, but we do have clues from cases in which these areas are damaged. The findings are disorders of the resting tone of muscles, difficulty starting movements, and unwanted involuntary movements. In Parkinson's disease, the input to the striatum from the substantia nigra is damaged. The result is a massive and disabling increase in muscle tone, accompanied by tremor in the hands, and great difficulty in starting movements. A very severe form of damage to the striatum is seen in late-stage Huntington's disease, where the face, trunk, and limbs writhe continuously. In the case of

The internal capsule

The internal capsule is a very large fiber sheet that lies between the putamen and globus pallidus on one side, and the thalamus on the other. All the fibers connecting the cerebral cortex with the thalamus, brainstem and spinal cord pass through the internal capsule, so a cerebral hemorrhage in the internal capsule can be devastating. The caudal continuation of the internal capsule is called the cerebral peduncle.

cerebral palsy, it is likely that the striatum and pallidum are damaged in the baby before it is born, and the result is overwhelming disorders of tone (often called spasticity) and a variety of involuntary movements.

The final common pathway for all motor systems – the motor neuron

All motor control systems act through alpha motor neurons which are the final pathway to muscles. These motor neurons in the ventral horn of the spinal cord activate the skeletal muscles of the limbs and trunk. Similar neurons in the midbrain and hindbrain activate the voluntary muscles of the head. Each alpha motor neuron is connected to many fibers in a skeletal muscle. When the alpha motor neuron fires, all of the fibers to which it is connected will contract. The combination of a single alpha motor neuron and the muscle fibers to which it connects is called a motor unit. Activating alpha motor neurons is the only way to cause skeletal muscle fibers to contract. As a result, all the motor control systems described above achieve their effects by stimulating or suppressing alpha motor neurons. The English physiologist and Nobel Prize winner, Charles Sherrington, described the role of alpha motor neurons as the final common pathway of the motor system.

Some incoming stimuli activate local interneuron networks to initiate patterns of activity called reflexes. These networks make rapid, corrective adjustments to muscle activity, to maintain muscle length, prevent overloads, imbalances and stresses, and avoid damage. The specific reflex response depends on the pattern of sensations that triggered it. Many spinal reflexes are driven by activity from sensory structures embedded in muscles. These are the muscle spindles, adjustable and very sensitive length-detectors, and the Golgi tendon organ, a force sensor located in tendons that attach muscles to a bones.

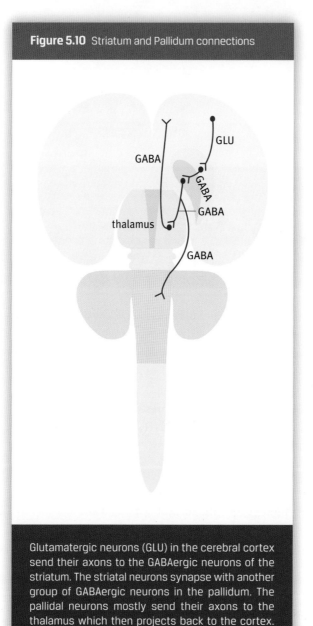

Figure 5.10 Striatum and Pallidum connections

Glutamatergic neurons (GLU) in the cerebral cortex send their axons to the GABAergic neurons of the striatum. The striatal neurons synapse with another group of GABAergic neurons in the pallidum. The pallidal neurons mostly send their axons to the thalamus which then projects back to the cortex. (Adapted from Swanson, 2007, p 177, flat map adapted from Swanson, 2004)

Curare and neuromuscular synapses

Curare is a plant extract prepared by some South American Indian tribes for use as an arrow poison for hunting. Curare blocks the effects of the neurotransmitter acetylcholine at neuromuscular synapses, and so paralyses the muscles. Because curare is only active when injected, the meat from the paralyzed animal can be eaten without danger. Curare analogs are used to relax muscles during many types of surgery. Because they paralyze the respiratory muscles, the anesthetized patient must have breathing assisted by a mechanical ventilator.

When a muscle is stretched, muscle spindle receptors are activated. The spindle receptors send a signal that stimulates the alpha motor neurons supplying the stretched muscle, making it increase tension to adjust to the new load. This reflex occurs via a single excitatory synapse between the stretch receptor axon and the dendrites of the motor neuron. This is a fast pathway that is called the monosynaptic stretch reflex. To keep reflexes active while the muscle changes length during movement, the sensitivity of muscle spindles is adjusted by small (gamma) motor neurons controlling muscle fibers inside the spindles. The gamma motor neurons are found in the ventral horn of the spinal cord, but are much smaller than the alpha motor neurons.

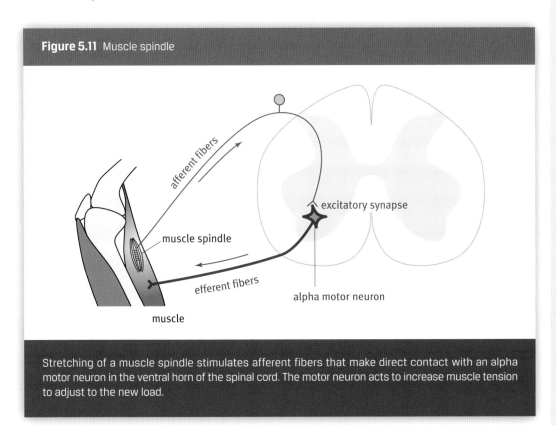

Figure 5.11 Muscle spindle

Stretching of a muscle spindle stimulates afferent fibers that make direct contact with an alpha motor neuron in the ventral horn of the spinal cord. The motor neuron acts to increase muscle tension to adjust to the new load.

Testing stretch reflexes

One of the standard parts of a neurological examination is the testing of stretch reflexes. This is done by tapping a tendon (such as the patellar tendon just below the knee) with a rubber-tipped hammer. An unusually brisk stretch reflex is often found in individuals who have suffered a stroke, whereas the stretch reflex is weak or absent if the relevant spinal nerves have been damaged.

When a muscle develops too much force for safety, Golgi tendon sensors are stimulated. They can prevent overload by inhibiting the alpha motor neurons of that muscle. In this way, the Golgi tendon sensors trigger an overload cut-out to reduce the force to a safe level.

Even these simple reflexes also have effects on antagonist muscles. For example, a stretch reflex that stimulates biceps neurons also inhibits triceps neurons. These reflex patterns can be combined to make large-scale corrections. For example, rolling your ankle will trigger a stretch reflex to straighten the ankle, but that is only part of the large postural correction reflex activity triggered by the same stimulus in order to maintain balance.

Reflexes are very useful for automatic, rapid responses, but can be counter-productive if they get triggered accidentally. Hence the other motor control systems normally suppress reflexes slightly, releasing them to be ready for action whenever we initiate a voluntary movement. People with spinal injuries lack the inhibitory connections coming down the spinal cord, so they have exaggerated reflexes.

Autonomic nerves supplying the heart

The heart rate is controlled by a balance of the action of sympathetic and parasympathetic nerves. The sympathetic nerves cause the heart to speed up, and the parasympathetic nerves make it slow down. If all the nerves to the heart are cut, as occurs in a heart transplant, it continues to beat at about 110 beats per minute, because it has an intrinsic pacemaker system. Under resting conditions, the parasympathetic nerves slow the heart to about 70 beats per minute.

Adrenaline and noradrenaline

These two transmitter molecules are very similar in structure, and have many actions in common. Almost all of the noradrenaline in the brain is manufactured in the locus coeruleus of the hindbrain, whereas adrenaline is manufactured in the adrenal medulla. Adrenaline is also called epinephrine, and noradrenaline is called norepinephrine.

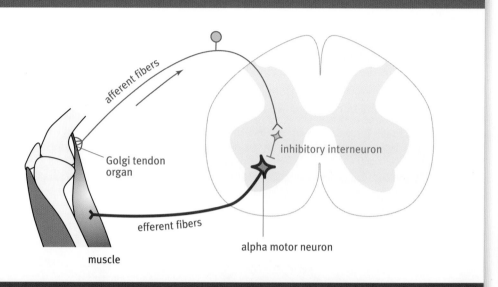

Figure 5.12 Golgi tendon sensors

The Golgi tendon organ contains sensors which respond to dangerous levels of muscle stretch. The Golgi tendon sensors stimulate afferent fibers that activate an inhibitory interneuron. The interneuron inhibits the motor neurons which supply the over-stretched muscle.

Command and control of the viscera – the autonomic nervous system

The effector systems that control the internal organs, glands, and blood vessels are collectively called the autonomic motor system, because most of their functions are automatic and involuntary. They can also be called the visceral motor system, as distinct from the somatic motor system that controls the voluntary muscles. The autonomic motor systems can in turn be divided into sympathetic and parasympathetic groups. The sympathetic group is involved in emergency (fight or flight) responses, whereas the parasympathetic group controls the normal day-to-day functions of the internal organs, such as salivation and digestion. Finally, the local neuron networks that coordinate the actions of the stomach and intestines are often grouped together under the name enteric nervous system.

Location of the sympathetic and parasympathetic motor centers

The motor neurons that supply the voluntary muscles are found in all regions of the brainstem and spinal cord, whereas the autonomic motor centers occupy only restricted areas. These include the sympathetic motor centers that are only found in thoracic and upper lumbar spinal cord segments (T1 to L2) and the parasympathetic motor centers that are only found in the brainstem and in the sacral spinal cord. The cervical spinal cord, the coccygeal spinal cord, and most of the lumbar spinal cord contain neither sympathetic nor parasympathetic motor neurons. The restricted origin of the sympathetic and parasympathetic motor systems poses a challenge for distribution, since each system must supply the entire range of viscera, and the sympathetic efferents also need to reach the skin and blood vessels of the whole body. The sympathetic and parasympathetic systems have developed different solutions to enable them to distribute their activity.

Whereas the somatic motor outflow uses a single motor neuron to connect the brainstem or spinal cord to a target muscle, the autonomic motor outflow is formed by a two-neuron chain. The first autonomic motor neuron in the chain is located inside the brainstem or spinal cord, and sends its axon to a second effector neuron located in a ganglion outside of the central nervous system. The axon of the second neuron then innervates the target organ. The two-neuron system enables a single autonomic motor neuron to trigger hundreds of ganglionic effector neurons, producing widespread effects, such as capillary constriction to raise blood pressure. By contrast, somatic motor neurons control small groups of fibers within a single muscle, and can be independently activated.

Because the second motor neurons in the autonomic chain are clustered together in ganglia, the first neuron is called preganglionic and the second neuron is called postganglionic. In the case of the sympathetic nervous system, the motor ganglia are close to the spinal cord, so the preganglionic cells have only short axons. In the case of the parasympathetic nervous system, the motor ganglion cells are as close as possible to the target organs (even embedded in their walls), and so the axons of the preganglionic cells are often relatively long.

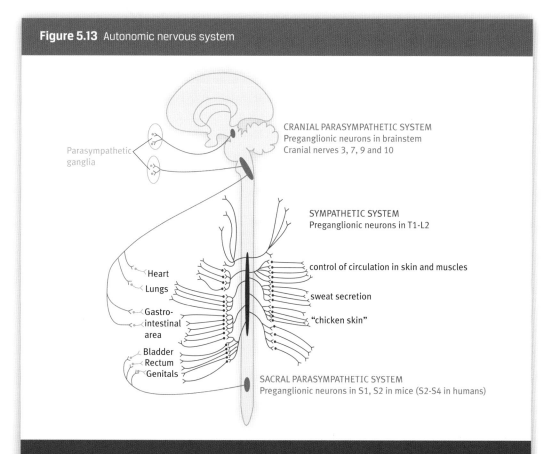

Figure 5.13 Autonomic nervous system

This diagram shows the origin and course of the sympathetic preganglionic neurons (colored red) and parasympathetic preganglionic neurons (colored blue). The postganglionic neurons are colored pink and pale blue. The parasympathetic preganglionic neurons are located in the midbrain and hindbrain, and in the sacral spinal cord. The sympathetic preganglionic neurons are located in the thoracic and upper lumbar spinal cord. (Adapted from Brodal, 2004, p 372)

The adrenal gland

The adrenal gland sits on top of the kidney. It is a combination of two separate endocrine glands. The center (medulla) of the adrenal gland manufactures and releases adrenaline under the control of the sympathetic nervous system. The outer layer (the adrenal cortex) manufactures and releases cortisol and related hormones under the control of the hypothalamus and the pituitary gland.

Functions of the sympathetic nervous system

The sympathetic nervous system is best thought of as the fight or flight system. It rapidly mobilizes key bodily systems in times of emergency or stress, preparing the body for action by raising blood pressure, relaxing and opening airways, mobilizing energy sources, and temporarily shutting down digestion. Most importantly, it increases blood flow to muscles by shutting down blood flow to the skin and intestines.

The sympathetic nervous system can increase its influence over the blood vessels of the body by triggering the release of adrenaline and noradrenaline from the adrenal medulla. These two hormones have essentially the same effect as the postganglionic sympathetic nerves, but because they remain in the blood stream, they amplify and prolong the sympathetic actions, and can influence cells that are not directly innervated by sympathetic nerves.

Sympathetic distribution

The preganglionic sympathetic neurons are found only in the middle region of the spinal cord, from the second thoracic segment to the first lumbar segment. Distribution to more rostral and caudal areas of the body is achieved by the existence of the sympathetic chain–a long system of ganglia and nerves that runs alongside the bodies of the vertebrae. The sympathetic motor ganglia are embedded in this chain with approximately twenty ganglia on each side. From the ganglia, postganglionic sympathetic fibers are distributed to all levels of the head and trunk, and also to the limbs.

Functions of the parasympathetic nervous system

The parasympathetic nervous system is responsible for the normal day-to-day functions of the internal organs (a role which can be described as 'rest and digest'). Parasympathetic actions on the cardiovascular system are opposite to those of the sympathetic nervous system; the parasympathetic outflow lowers the heart rate and blood pressure. The parasympathetic nervous system does not send nerves to the limbs, so its role is restricted to the glands and organs of the head and trunk. It is also responsible for penile erection and bladder emptying.

Parasympathetic distribution

In the case of the parasympathetic motor system, there is a huge gap between the brainstem (cranial) output and the sacral output. A single nerve, the vagus nerve, fills the gap. The vagus nerve (cranial nerve 10) conveys preganglionic parasympathetic axons from the brainstem and travels from the base of the skull to the pelvic brim. In between, it supplies the larynx, esophagus, lungs, heart, stomach, and the small and large intestines. The name of the vagus is taken from a latin word meaning 'wanderer.' The sacral preganglionic neurons supply effector neurons in the sigmoid colon, rectum, bladder and genitalia.

Transmitter substances in the autonomic nervous system

All sympathetic and parasympathetic preganglionic motor neurons are like somatic motor neurons in that they use acetylcholine as their transmitter substance. However, the sympathetic ganglionic effector neurons are mostly noradrenergic whereas the parasympathetic ganglionic motor neurons are cholinergic. Autonomic receptors are all metabotropic, meaning that they can be coupled to a range of intracellular mechanisms. In addition, the receptors which bind noradrenaline are of different kinds. Acetylcholine receptors also vary in terms of function. This variation in receptor types creates an even greater diversity of effects. For example, the

---

Opponents or partners?

It is sometimes suggested that the sympathetic and parasympathetic nervous systems act in opposition to each other. It is true that in many parts of the body the autonomic effector neurons of each system actually inhibit the actions of the other. However, acetylcholine and noradrenaline do not interfere with each other, and receptors for both can be activated simultaneously. It is more accurate to think of the sympathetic and para-sympathetic systems as complementary to each other.

sympathetic neurons that secrete noradrenaline trigger different responses in the smooth muscle cells of the gut because of receptor differences; the same neurotransmitter causes the gut walls to relax on the one hand, and the gut sphincters to contract on the other, because of their different adrenergic receptors.

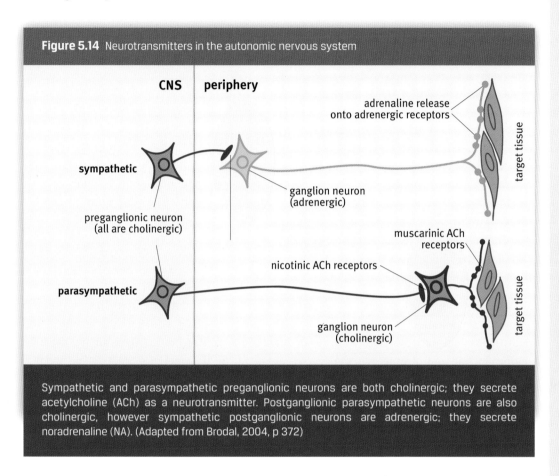

Figure 5.14 Neurotransmitters in the autonomic nervous system

Sympathetic and parasympathetic preganglionic neurons are both cholinergic; they secrete acetylcholine (ACh) as a neurotransmitter. Postganglionic parasympathetic neurons are also cholinergic, however sympathetic postganglionic neurons are adrenergic; they secrete noradrenaline (NA). (Adapted from Brodal, 2004, p 372)

Enteric mysteries

The enteric nervous system has proven remarkably difficult to study, since it is spread across the entire length of the gut and is embedded between layers of muscle and connective tissue. In addition, the familiar categories of sensory neuron, interneuron, and motor neuron are blurred and intermingled, with many enteric neurons performing two or even all three of these functions.

Higher autonomic control

The motor neurons of the autonomic nervous system are regulated by centers in the brainstem that control vital functions of the cardiovascular and respiratory systems, as well as salivation, swallowing, digestion, peristalsis, defecation, and urination. At a level above the brainstem, the hypothalamus triggers these autonomic functional centers as part of larger behavioral responses, such as those involved in aggression. Anger is associated with increases in heart rate, blood pressure, and respiratory rate.

The enteric nervous system

The networks of neurons embedded in the wall of the gut, which regulate and coordinate its activity, are called the enteric nervous system. This subdivision of the autonomic nervous system deserves a special name because it is so large. In humans, it contains 200-600 million neurons, more than in the entire spinal cord. The enteric nervous system controls the contractions and secretions of the stomach, intestines, and other digestive organs. Although it is influenced by the central autonomic nervous system, it is capable of independent action through its own reflex systems. The enteric nervous system is also functionally linked to the immune system of the body, to help defend against invasive microorganisms in the gut. In many ways, the enteric nervous system seems to be evolutionarily older than the central nervous system, presumably because food intake and digestion is a priority for simple organisms.

The enteric nervous system makes use of about twenty different transmitter substances, the most important of which are acetylcholine, vasoactive intestinal polypeptide (VIP), nitric oxide (NO), GABA, and serotonin. Complex networks in the enteric nervous system can learn to adapt to challenges in the absence of CNS control. In Hirschsprung's disease, there is a congenital absence of parasympathetic innervation of the sigmoid colon, which makes the sufferer unable to defecate. Treatment consists of removal of the non-innervated region of colon. About a year after removal of the denervated section, the remaining colon learns to defecate.

Command and control of the neuroendocrine system

The definition of a motor neuron is a nerve cell whose cell body lies within the central nervous system, but whose axon travels out of the central nervous system to initiate some action. In most cases the actions are achieved through the contraction of muscles. However, many autonomic motor neurons do not act on muscles but on glands, such as the salivary glands, which are made to produce secretions. Some sympathetic motor neurons act by releasing adrenaline from the medulla of the adrenal gland. The adrenaline is secreted directly into the blood stream. This type of action is described as neuroendocrine.

The hypothalamus has a number of powerful neuroendocrine motor neurons that control the hormonal secretions of the pituitary gland. The pituitary gland is the master gland of the endocrine system, once described as 'the leader of the endocrine orchestra.' There are two separate neuroendocrine control systems in the hypothalamus, one for the anterior pituitary gland and one for the posterior pituitary gland. The hypothalamic neurons that control the anterior pituitary are small (parvicellular), whereas those that send hormones to the posterior pituitary are large (magnocellular) neurons.

The anterior lobe of the pituitary is under the control of parvicellular neurosecretory cells of the hypothalamus. These neurons do not extend their axons to meet their targets but instead communicate with target cells via the bloodstream. They do this by releasing control hormones (hypophysiotropic hormones) into a unique capillary system called a portal blood system. This portal blood system then transports the hypophysiotropic hormones downstream where they meet and bind to specific pituitary receptors. This binding causes either stimulation or inhibition of anterior pituitary hormone release. Hormones released in the anterior pituitary are follicle-stimulating hormone (FSH), luteinizing hormone (LH), adrenocorticotrophic hormone (ACTH), thyroid-stimulating hormone (TSH), growth hormone (GH) and prolactin (PRL). These hormones are capable of exerting powerful effects on growth, reproduction and regulation of the body.

The hypothalamic cell groups that connect with the posterior pituitary are the paraventricular and supraoptic nuclei–the magnocellular neurosecretory nuclei of the hypothalamus. Together they manufacture substantial amounts of two important hormones, oxytocin and vasopressin. The cells that manufacture oxytocin and vasopressin are very large and stain deeply with Nissl stains. The dark staining material is the rough endoplasmic reticulum in these very busy cells. Oxytocin and vasopressin travel along the axons of these cells to reach the posterior lobe of the pituitary, where they are released into the blood stream.

Hypothalamic headquarters

The hypothalamus is able to activate responses in somatic, visceral, and endocrine effectors, based on incoming information about bodily state and physiological needs. Nuclei in the hypothalamus regulate sleep, arousal, feeding, drinking, reproduction, and rivalry, as well as motivating the cerebral cortex to find behavioral solutions to bodily needs, such as strategies to find food or water. Hypothalamic nuclei coordinate most of the core survival capabilities of the nervous system.

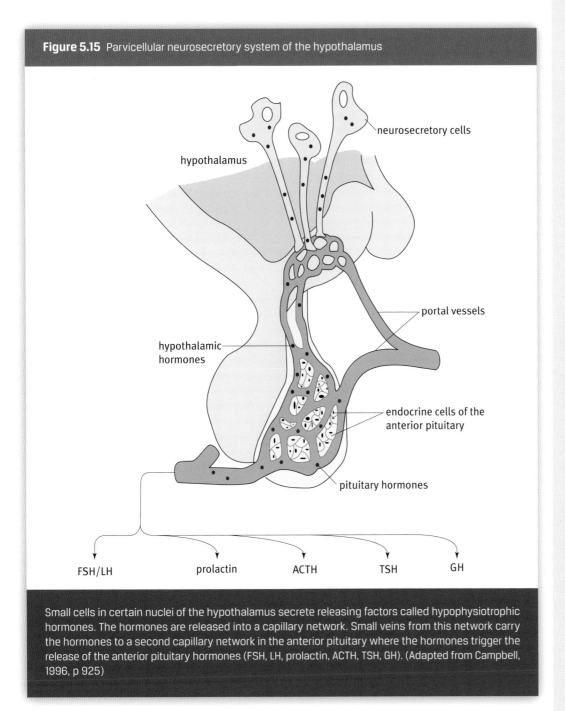

Figure 5.15 Parvicellular neurosecretory system of the hypothalamus

Small cells in certain nuclei of the hypothalamus secrete releasing factors called hypophysiotrophic hormones. The hormones are released into a capillary network. Small veins from this network carry the hormones to a second capillary network in the anterior pituitary where the hormones trigger the release of the anterior pituitary hormones (FSH, LH, prolactin, ACTH, TSH, GH). (Adapted from Campbell, 1996, p 925)

Portal blood systems

In most parts of the body, arteries supply a capillary bed, and veins collect the capillary blood. In a portal system, there are two separate capillary beds, connected to each other by small veins. This allows products collected by the first capillary bed to be transported exclusively to the second capillary bed for release into local tissues. In humans, there are two portal systems, a very large one carrying nutrients from the intestines to the liver, and a very small one carrying hormones from the hypothalamus to the pituitary.

Oxytocin has a wide range of functions relating to birth and the care of offspring. At the end of pregnancy, oxytocin initiates contraction of the uterus, which causes the baby to be born. Following the birth, oxytocin stimulates the production of breast milk. Suckling of the nipple by the baby causes more oxytocin secretion, and so prolongs the period of lactation. In recent years it has been found that oxytocin also has profound effects on maternal behavior. Many parts of the brain have receptors that respond to oxytocin secreted by the paraventricular nucleus. Oxytocin causes the mother of a newborn baby to feel loving and protective, and to become psychologically attached to the baby. It seems that oxytocin may also help to keep mothers and fathers together by promoting good feelings about the relationship. It is likely that oxytocin is released after orgasm, so that it helps keep sexual partners together.

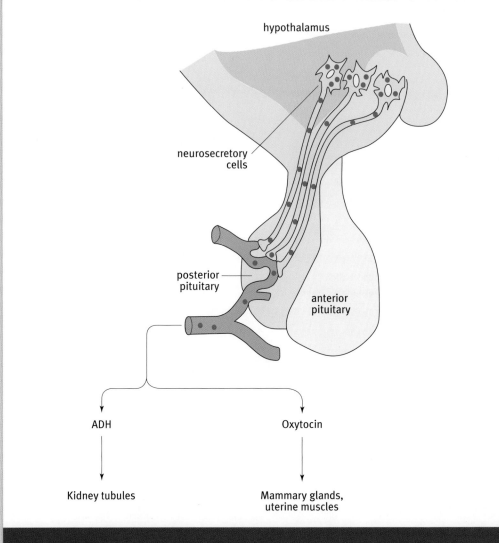

Figure 5.16 Magnocellular neurosecretory system of the hypothalamus

Large cells in the paraventricular and supraoptic nuclei of the hypothalamus secrete oxytocin and ADH (vasopressin). These neurotransmitters travel to the posterior pituitary where they are released into the blood stream, and so become hormones. (Adapted from Campbell, 1996, p 925)

Paraventricular and supraoptic nuclei

These two hypothalamic nuclei manufacture the posterior pituitary hormones oxytocin and vasopressin. In Nissl-stained sections they are both stained very darkly. The reason for this is the large amounts of active rough endoplasmic reticulum (RER) present in their neurons. RER is the protein-manufacturing component of cells. RER is particularly well stained with the Nissl method. Very active manufacturing cells, like those of the paraventricular and supraoptic nuclei, are therefore prominent in Nissl-stained sections.

Vasopressin (also called antidiuretic hormone—ADH) has powerful effects on blood pressure and on urine production. It can raise blood pressure, and can minimize loss of water in the urine in dry conditions. It also has functions similar to those of oxytocin.

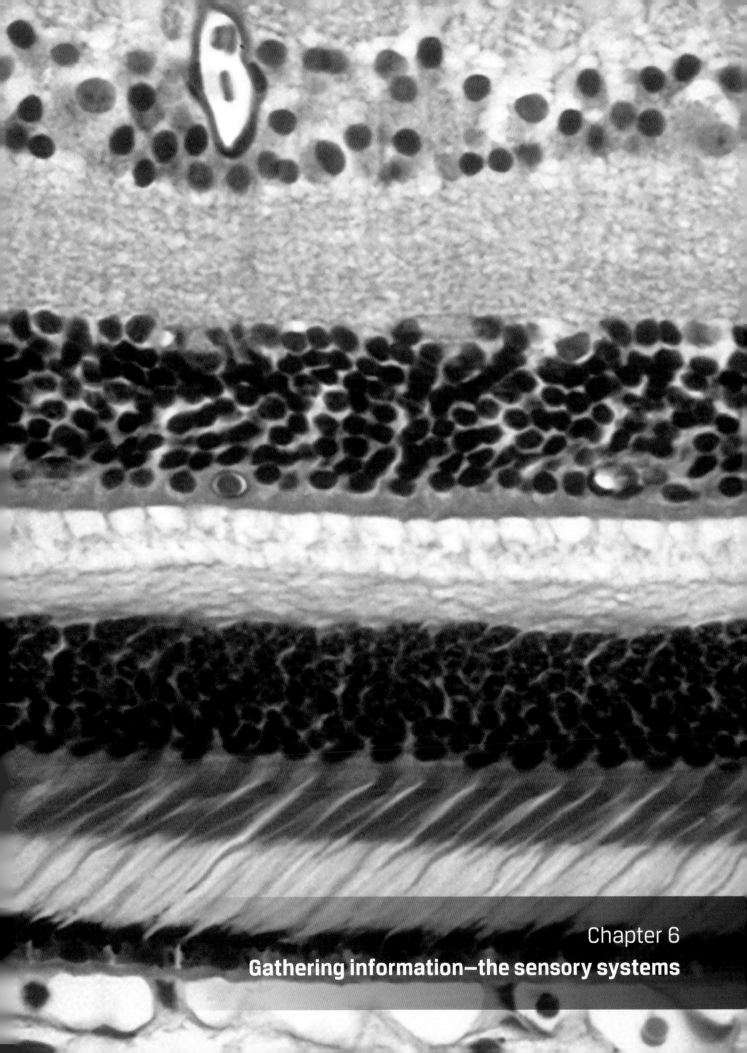

Chapter 6
Gathering information—the sensory systems

Gathering information —the sensory systems

In this chapter:

Receptors	76
Keeping sensory maps intact	77
Interpretation and understanding	77
Sensory areas in the cerebral cortex	78
Vision	84
Hearing	87
Vestibular system	89
Taste	93
Smell	93
Sensory processing outside the cortex	95
An example: rolling an ankle	95

The nervous system controls behavior and bodily states in order to best respond to the challenges of the environment. To do this well, it needs information about the external world and the internal state of the body. The parts of the nervous system that receive and process information are termed sensory systems.

There are five senses common to most mammals; vision, smell, hearing, taste, and touch. However, some mammals have developed other special systems that aid in their survival. The best examples are echolocation (in bats and dolphins) and electroreception (in the platypus). Some mammals, such as the mole, have lost almost all their visual sense, and rely almost entirely on smell and touch.

The sensory systems have two basic roles–detection and understanding. Detection refers to the recognition of information, while understanding refers to using this information appropriately. Detection is the role of specialized receptor cells; the highest level of understanding occurs in the cerebral cortex.

When we speak of the 'senses,' we are generally referring to conscious perception, but this is not the only way that sensory information is used by the central nervous system. Conscious perception is just one aspect of sensory function, since the information collected by sensory receptors is registered at many levels in the nervous system. In many cases, the conscious experience of sensation may occur well after other control centers have already acted on the information. This highlights the distinction between receptor activity, indicating that a receptor has been stimulated, and perception, which is the conscious experience that results from the cortex receiving information from the receptor activity.

Receptors

Receptors convert energy of a particular type (such as light, vibration, or pressure) into a graded electrical potential, which is transmitted to the central nervous system by a sensory axon. The process of converting a physical stimulus into a graded electrical potential is called transduction. Receptors use changes in the permeability of ion channels to alter their membrane potential. If the stimulus is strong enough, it will generate an action potential in the sensory axon. Each action potential in a sensory axon is identical, but stimuli of different intensity or duration can alter the rate of firing of the sensory axon, and so give more information about the stimulus. Each sensory modality employs an exclusive pathway to the cortex, making specific and ordered connections, so the brain can make sense of the information it receives.

Most vertebrates have similar types of receptors, but individual groups of mammals have specializations in terms of the exact type and number of receptors. Primates are largely visual animals and have many more visual receptors, including some specialized to detect color. Rats and mice, which usually live in dark and confined environments, have fewer visual receptors, but have many tactile receptors linked to whiskers, which sweep in front of them as they move around.

The electrical signal generated by the receptor is passed on to a ganglion cell, whose job is to carry the signal to the brain or spinal cord. In the case of fine touch sensation from a fingertip, this is the dorsal root ganglion neuron has a long axon that travels from the fingertip, past its ganglion cell body, and into the spinal cord. In the central nervous system, this same axon can travel rostrally and pass the signal on to a neuron in the hindbrain. The axon arising from this hindbrain neuron crosses to the opposite side and travels to the thalamus. This means that the left thalamus receives touch information from the right side of the body. The final link consists of a thalamic cell axon which sends the information to the cerebral cortex.

This sequence of information transmission, from receptor, ganglion cell, relay neuron, thalamic neuron, and finally to cortex, is present in almost all sensory systems. The basic plan does have some important variations; not all the information is sent to the opposite side of the nervous system. In the case of the auditory system there are both crossed and uncrossed pathways, which are interconnected in an elaborate system for detecting the exact direction from which the sound came.

Keeping sensory maps intact

A common property of sensory projections is that they retain some kind of spatial arrangement for their entire passage through the nervous system. The maintenance of spatial (topographical) arrangement results in what is called somatotopy for the touch system, which means that the map of the skin of the body in proper spatial order is transmitted to the cortex. In the case of the visual system, the retinal map is maintained (retinotopic organization) and in the case of the auditory system, the tonal map is transmitted intact (tonotopic organization).

Interpretation and understanding

Understanding involves the interpretation of sensory information, so that the brain can take appropriate action. At the simplest level, sensory stimuli may provoke an immediate reflex response. At the next level, sensory interpretation might include making use of sensory feedback to modify a pattern of activity, such as altering walking behavior on the basis of feedback from

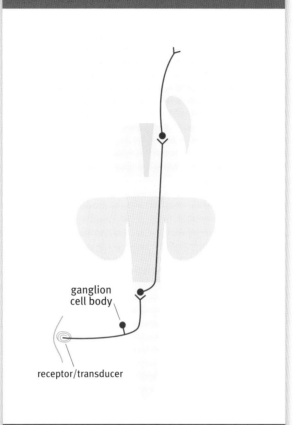

Figure 6.1 The three neuron sensory pathway from periphery to cortex

This diagram shows the basic plan of a sensory pathway that ends in the cortex. Information from a receptor (light, sound, touch, or taste) is picked up by the primary sensory neuron. The cell body of the primary sensory neuron lies outside the central nervous system but its axon connects with sensory nuclei that project to the thalamus. The projection to the thalamus generally crosses the midline. The thalamus sends the information to a specific cortical sensory area. (Flat map adapted from Swanson, 2004)

The thalamus

The thalamus is a paired structure lying next to the third ventricle. All the sensory pathways, except olfaction, synapse in the thalamus. The thalamic nuclei then project to the various sensory areas of the cerebral cortex. Thalamic nuclei also project to areas of cortex outside of the sensory areas, such as the prefrontal cortex and the parieto-temporal association areas.

the soles of the feet and the angles of leg joints. At a more conscious level, understanding might include using the cerebral cortex to decode the sounds of speech, and using the speech content as part of a decision making process for action. The essence of understanding is the interpretation of sensory information on the basis of prior learning, in order to select the most useful responses.

Sensory areas in the cerebral cortex

Each of the senses is received by a separate area of cerebral cortex. Skin sensation is received by the primary somatosensory cortex (S1), which lies immediately behind the motor cortex. The primary visual cortex (V1) is located more caudally in the occipital pole of the cortex. The primary auditory cortex (A1) is located in the temporal region. In the human brain, the A1 cortex is mostly hidden in the lower bank of the lateral fissure, and the V1 cortex is mostly located on the medial side of the occipital lobe. Visceral sensation (including taste and sensation from internal organs) is received by the insula, deep in the lateral fissure.

In each of the primary cortical sensory areas the representation is spatially organized. For example, the S1 cortex contains a map of the whole skin surface of the opposite side of the body, with the parts of the body properly arranged. In the V1 cortex, there is a precise point-to-point map of the visual field that is seen by the eye. Although these cortical maps are spatially coherent, they are distorted in an interesting way. Some parts are represented by a much bigger cortical area than others. In humans the areas receiving touch information from the lips and fingertips are huge compared with the area devoted to the rest of the body.

The insula

The insula is an area of cerebral cortex that receives taste sensation and information from the stomach and intestines. It is located adjacent to the mouth area of the somatosensory cortex. In the human brain, the insula is hidden by the overgrowth of the banks of the lateral fissure. The name 'insula' means an island, in reference to the isolated position of this small area of cortex. In rodents this region is on the surface, but is still called the insular cortex for consistency.

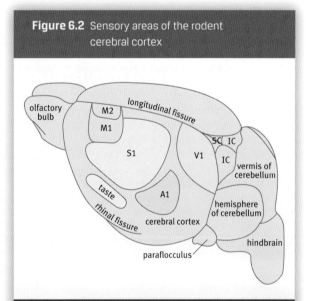

Figure 6.2 Sensory areas of the rodent cerebral cortex

All mammals have well defined areas in the neocortex for touch (somatosensory-S1), vision (V1), hearing (A1), and taste. S1 is always the most rostral, lying next to the motor areas (M1 and M2). V1 is always the most caudal, located in the occipital pole. A1 is ventrally placed between S1 and V1. The taste area is ventral to S1. The olfactory bulbs do not project to the neocortex; the olfactory cortical areas are three layered regions on the ventrolateral surface of the brain below the rhinal fissure.

Each area of primary sensory cortex is surrounded by secondary areas that progressively analyze and interpret sensory information. For example, the image received by the visual cortex may be recognized in a related cortical area as the face of someone we know. Finally, different kinds of sensory information (touch, vision, hearing) are pooled in association cortical areas in the parietal and temporal lobes. This enables the brain to correlate what we see with what we hear and feel.

It is important to note that the olfactory pathway has a completely different organization. The olfactory pathway does not cross to the opposite side of the brain as the touch pathway does, and the thalamus is not involved in the chain from periphery to cerebral cortex. Instead, the olfactory pathways go directly to the olfactory cerebral cortex, an area near the amygdala called the piriform cortex.

Somatosensory system

The somatosensory system carries information about touch, deep pressure, pain, and temperature from the skin, joints and muscles. There are two major somatosensory pathways, the dorsal column-medial lemniscus pathway and the spinothalamic pathway. Each consists of three neurons forming a chain from the receptors to the cerebral cortex. The primary sensory neuron has its cell body in a spinal ganglion or cranial nerve ganglion, the secondary sensory neuron has its cell body in the gray matter of the spinal cord or in the hindbrain, and the tertiary sensory neuron has its cell body in the thalamus. These pathways cross the midline so that the sensory information from one side of the body is brought to the opposite cerebral hemisphere. A major difference between the two pathways is the levels at which they cross the midline; temperature and pain-related fibers (spinothalamic tract) cross almost immediately where they enter the spinal cord, whereas fine touch and proproception (dorsal column-medial lemniscus pathway) does not cross until it reaches the hindbrain.

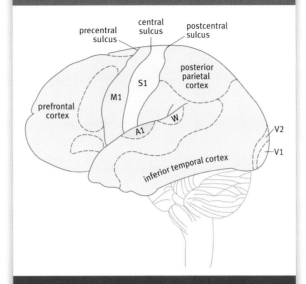

Figure 6.3 Sensory areas of the human cerebral cortex

As in rodents, the somatosensory cortex (S1) is located just caudal to the primary motor cortex (M1). The visual cortical areas (V1 and V2) are found in the occipital pole, but in the human are almost entirely on the medial side of the occipital lobe. The auditory cortex (A1) is located in the upper border of the temporal lobe just below S1. However, A1 is shown to lie on the surface of the brain in most textbook diagrams, most of A1 actually lies in the lower bank of the lateral fissure and is hidden from view. The taste cortex is also ventral to S1, but cannot be seen on the lateral surface because it is hidden in the depths of the lateral fissure. The taste cortical area in humans is called the insula.

Dorsal column-medial lemniscus pathway

This is the main touch pathway. It transmits signals from low-threshold mechanoreceptors in the skin, muscles, and joints. This enables tactile discrimination, vibration detection, form recognition, and proprioception (sense of position from joint and muscle receptors).

First-order (primary) neurons are located in the dorsal root ganglia. In the lower extremities, the axons of these ganglion cells form the gracile fasciculus, while in the upper extremities they form the cuneate fasciculus. These fasciculi are collectively known as the dorsal columns. The dorsal columns are somatotopically organized, with the gracile axons placed medially, and the cuneate axons lateral to them. The axons ascend in the dorsal columns and terminate in the gracile and cuneate nuclei of the hindbrain. This is where the second-order neurons are located. The axons of these neurons cross the midline and form a compact bundle called the medial lemniscus. It ascends through the brainstem to synapse with third order neurons in the ventroposterior nucleus of the thalamus. The thalamic neurons send their axons to the somatosensory cortex in the postcentral gyrus.

Loss of the dorsal column system

A hundred years ago there were two diseases that commonly destroyed the dorsal column system, tertiary syphilis and vitamin B12 deficiency (which was then called pernicious anemia). The old name given to loss of the dorsal columns was tabes dorsalis, which means a wasting of the dorsal columns. Dorsal column loss is associated with loss of touch, deep pressure sense, and position sense. The discovery of penicillin caused a dramatic reduction in the number of cases of tertiary syphilis, and the discovery of vitamin B12 made it possible to treat pernicious anemia. Tabes dorsalis is very rare today.

Crossed and uncrossed pathways

For unknown reasons, the cerebral cortex of one side is generally connected with the opposite side of the body. Because of this, sensory and motor pathways connected to the cerebral cortex must cross the midline at some point. The opposite is true for each side of the cerebellum, which is consistently related to the same side of the body. Connections between the cerebral cortex and the cerebellum must therefore cross the midline to maintain consistency.

[5] This sensory system is not present in the human nervous system

Table 6.1 A Comparison of Different Sensory Systems

Sense	Receptors	Nerves and tracts	Subcortical relay centers	Cortical areas
vision	rods, cones	optic nerve and optic tract	lateral geniculate, superior colliculus	visual cortex in the occipital lobe
hearing	hair cells	vestibulocochlear nerve	cochlear nuclei, superior olive, inferior colliculus, medial geniculate	auditory cortex in lower bank of the lateral fissure in the temporal lobe
skin senses	Pacinian, Merkel, and Ruffini receptors, free nerve endings	spinal nerves and the trigeminal nerve	dorsal spinal cord, hindbrain nuclei, and thalamic nuclei	somatosensory cortex–postcentral gyrus
proprioception	vestibular, spindles, joint capsules	spinal nerves, cranial nerves	dorsal spinal cord, hindbrain and thalamic nuclei	somatosensory cortex
smell	olfactory receptors	olfactory bulb and tract	none	piriform cortex
taste	taste buds	facial and glossopharyngeal nerves	nucleus of the solitary tract and gustatory thalamus	insular cortex
pheromone detection[5]	special olfactory receptors	accessory olfactory bulb and tract	none	bed nucleus of the accessory olfactory tract
internal senses	stretch receptors, free endings	visceral afferents	nucleus of the solitary tract	insular cortex

Spinothalamic tracts

There are two spinothalamic tracts, the lateral tract and the ventral tract. Information about pain and temperature is conveyed by the lateral spinothalamic tract while touch sensation is conveyed by the ventral spinothalamic tract. Axons connect the skin with the dorsal horn of the spinal cord, where they synapse almost immediately with second order neurons in the substantia gelatinosa, a loose aggregate of neurons at the head of the dorsal horn. The cell bodies of neurons that make up the spinothalamic tract are located principally within the dorsal horn of the spinal cord, and their axons cross the midline via the anterior white commissure of the spinal cord. The axons then continue along the length of the spinal cord toward the thalamus, where they ultimately synapse with third-order neurons, which in turn transmit signals to the somatosensory cortex.

Pain and nociception

It is important to distinguish between pain, a state of distress usually related to injury or damage, and nociception, the detection of actual or likely tissue damage. Pain and nociception are closely related, but can occur independently. A person under general anesthetic has working nociceptors

Figure 6.4 Somatotopic organization of the postcentral gyrus

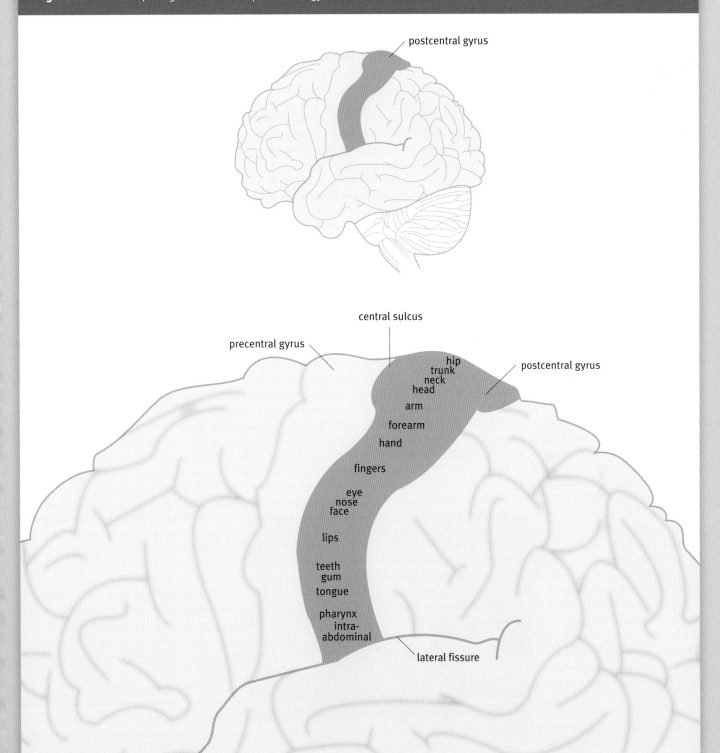

Skin sensation is represented on the postcentral gyrus in the form of an upside down and distorted body map. The lower limb is represented most dorsally and the face and tongue are represented most ventrally, near the lateral fissure. The size of the representation of each part of the body is proportional to its importance in touch sensation. The fingertips and lips occupy large areas compared to the lower limb and trunk. (Adapted from Bear et al, 2007, p 402)

that can signal tissue damage and trigger reflexes, but does not experience pain because there is no conscious awareness of distress to motivate behavior. Conversely, in some cases of chronic pain, the nervous system can end up abnormally sensitized so that the pain state persists even when the nociceptors are no longer active, a condition called neuropathic pain.

Pain is a signal that the body may be damaged, and the injury must be avoided as a high priority. Because pain is such an important signal to the nervous system, it is typically not subject to the same kinds of filtering that occur with other senses. For example, most senses will adapt to a stimulus that persists for a long time, and the system will respond to it less and less over time. However, if nociception persists for long periods, chemical feedback mechanisms make nociceptors become more sensitive and can cause previously unresponsive receptors to become activated (called peripheral sensitization). In CNS pathways carrying nociceptive information, chronic stimulation causes other interactions, which make the pathways more responsive to incoming stimuli (called central sensitization). Sensitized second-order neurons become hypersensitive to nociception, and may also signal pain when triggered by innocous stimuli, such as touch. Clinically this painful response to innocuous stimuli is called allodynia.

Spinal cordotomy

In some cases of intractable pain, the lateral spinothalamic tracts may be surgically cut to stop pain transmission. This is a drastic operation, but it has proved to be effective in patients with mesothelioma, an incurable cancer of the covering of the lung.

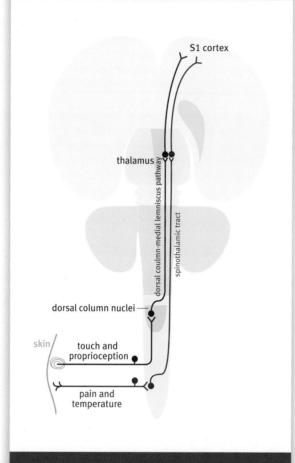

Figure 6.5 The somatosensory pathways from periphery to cerebral cortex

There are two somatosensory pathways, the dorsal column-medial lemniscus (DCML) pathway and the spinothalamic pathway. The DCML pathway carries fine touch and proprioceptive sensation and the spinothalamic pathway carries pain, temperature and touch sensation. The DCML pathway crosses the midline in the hindbrain whereas the spinothalamic fibers cross the midline at about the same level as the primary sensory neuron. Both these pathways project to somatosensory nuclei in the thalamus, which in turn project to S1. (Flat map adapted from Swanson, 2004)

Nociceptors

Nociceptors are triggered by actual or potential tissue damage. Generally, nociceptors are free nerve endings in tissue and respond to mechanical, thermal, or chemical stimuli. The sensory fibers that carry nociception are of two types–fast and slow. The fast nociceptor pathway is associated with sharp, well-localized pricking or stinging pain. These sensations travel via thin myelinated fibers (A delta fibers) to the superficial dorsal horn of spinal cord or, in the case of the head, to the caudal spinal trigeminal nucleus. The transmitter substance used is glutamate, which gives a rapid, brief transmission. The information travels in the spinothalamic tract to reach the ventroposterior thalamic nuclei, and is passed on to the somatosensory cortex.

The slow nociceptor pathway is associated with dull, burning, or aching sensations. These sensations travel via slower, unmyelinated fibers (C fibers) to deep layers of dorsal horn of the spinal cord or the caudal spinal trigeminal nucleus. The transmitter substance used by this pathway is substance P. The pathway projects via the spinothalamic tract to intralaminar nuclei of the thalamus, and is associated with poorly localized pain that is unusually distressing. As well as sending information to the cerebral cortex, nociceptor systems have a vital role in triggering reflexes that protect the body from the source of the pain, such as the withdrawal reflex in the limbs.

Ascending pathways

Nociceptors connect to dorsal horn and ascending projection neurons. These send axons across to the cortex via the thalamus. While the main touch pathways cross the midline in the hindbrain, pain pathways cross in the spinal cord just above the level where the nerves enter.

Visceral pain

We are most familiar with pain in our skin or joints, but pain from the internal organs is quite different. Visceral nociception is triggered by smooth muscle spasm, organ distension, inflammation, or ischemia (loss of blood supply). Because we are not usually conscious of visceral sensation, the brain finds it hard to localize the source of visceral pain. In some cases, the pain may seem to come from somewhere on the body surface. This phenomenon is called referred pain. The ache experienced in gall bladder inflammation is a good example. The gall bladder inflammation triggers visceral nociceptors, but instead of localizing the pain in the abdomen, the brain perceives the source to be a patch of skin on the shoulder. The reason for percieving pain in the shoulder is that both the shoulder and the coating of the diaphragm (near the gall bladder) are supplied by the C4 nerve. Visceral pain is often associated with feelings of distress, anxiety or depression, and can be extremely debilitating.

Headache

The brain itself has no nociceptors, but the blood vessels and meninges are very sensitive, and can cause pain in the head. Headache is at times caused by pain in a distant area, such as the neck or the teeth. In these cases, the pain is said to be referred to the head region.

The endogenous opiate system

The brain produces a number of chemicals that act like the opium derivatives, heroin and morphine. These chemicals (enkephalins and endorphins) are therefore called endogenous opiates. Endogenous opiates are used by the nervous system to control the transmission of nociception by reducing neurotransmission at synapses. They can also alter mood, which affects the perception of nociception as pain. These control mechanisms are essential at times when it is necessary to temporarily disregard nociception, such as in a life-threatening situation. Experiencing pain under those circumstances would distract the brain from organizing an escape behavior, so nociception is temporarily blocked off.

Proprioception

The term proprioception is constructed from the Latin word 'proprius' which means 'near'. Proprioception is therefore sense of nearness, or more broadly, sense of position. Individuals who suffer damage to the dorsal column system, such as in tertiary syphilis, show a loss of deep pressure sensation, which is normally important for proprioception. If such a person closes their eyes when standing up, they risk falling over because they have no sense of deep pressure from their legs.

The diaphragm and C4

The diaphragm and its coverings originally develop high up in the neck. They migrate down to the junction of thorax and abdomen, dragging their nerves (mainly from C4) behind them. This is the reason that pain on the surface of the diaphragm might be perceived in the lower neck or shoulder region. It also explains why individuals who suffer spinal cord transection in the lower neck can keep breathing.

Pain management

Many different drugs are used in pain management. Local anesthetics (such as lignocaine) disable nociceptors by preventing them from firing action potentials, while analgesics (such as aspirin, ibuprofen, and morphine) block pain signaling. General anesthetics (such as ether and halothane) are neither anesthetic nor analgesic, but they block pain perception by making the person unconscious.

Vision

The retina is the light-sensitive receptor layer at the back of the eye. Light passes through the cornea, the aqueous chamber, the lens, then the vitreous chamber of the eye to reach the retina. Both the cornea and the lens help to focus the image on the retina. The light receptor layer is at the very back of the retina, so light must pass through the surface layers of the retina before it reaches the receptor cells, the rods and cones.

The organization of the human retina

The human retina is a deep cup. Its shape represents about three-quarters of a sphere with a diameter of 22 mm. There are about 7 million cone cells and about 700 million rod cells in the human retina. The cone cells are concentrated in a central area called the macula, which is responsible for high acuity vision and color vision. There are very few rod cells in the macula. The rod cells are sensitive to very dim light, and react quickly to moving objects that enter the visual field. Cones are sensitive to bright light and fall into three categories that respond to light in the red, green or blue parts of the spectrum. Just medial to the macula is an area where the axons leaving the retina gather together to form the optic nerve. There are no photoreceptors in this area, so it forms a blind spot on the retina, about 1.5-2.0 mm in diameter. In rodents, as in all mammals except simian primates, there is no distinct macula. Instead, many mammals (especially herbivores) have a horizontal concentration of retinal cells called the visual streak. The visual streak seems to be useful for scanning the horizon in order to detect predators.

The retinal cell layers

The cells of the retina are arranged in distinct layers. From a functional point of view, the main three layers are the photoreceptor layer, the bipolar cell layer, and the ganglion cell layer. The photoreceptor layer is the deepest layer of the retina–furthest from the entry path of light rays. The receptor cells in this layer are the rods and cones. The light sensitive processes of the rods are partly embedded in the retinal pigmented epithelial layer (RPE). The exchange of retinoids between the RPE and photoreceptors is necessary for the photoreceptors to convert light signals into electrical signals. The bipolar cell layer contains bipolar cells, as well as amacrine and horizontal cells. The bipolar cells receive information from photoreceptors and send signals to ganglion cells. The role of the amacrine and horizontal cells is to modulate those signals. The ganglion cell layer contains large ganglion cells that have long axons; they transmit information received from bipolar cells to the brain, via the optic nerve. The ganglion cell axons reach centers in the thalamus (the lateral geniculate nucleus), hypothalamus (the suprachiasmatic nucleus), and the superior colliculus of the midbrain.

The networks of the retina are capable of significant signal processing and consolidation of information. For example, ganglion cells on the edges of the retina receive information from a hundred or more rod cells, thus increasing sensitivity to dim light. In contrast, ganglion cells in the macula of a human retina receive information from only two or three cones, increasing our capacity to resolve fine detail.

Color channels

Different animals have different numbers of visual pigments in their photoreceptors, which allows different types of color vision. Humans and related primates are trichromats, meaning that they have three pigments (sensitive to red, green and blue) and perceive color based on the different activation of these pigments. This is why red, green and blue elements in televisions and computer displays are able to simulate nearly the entire range of colors we can see. Other animals such as dogs and mice are dichromats, meaning that they have two color sensitive pigments and can see a range of colors limited by those pigments. Lampreys, which need to detect changes in water composition in order to breed, have five types of photoreceptors. Television for lampreys would need five color elements to look convincing.

Figure 6.6 The eye

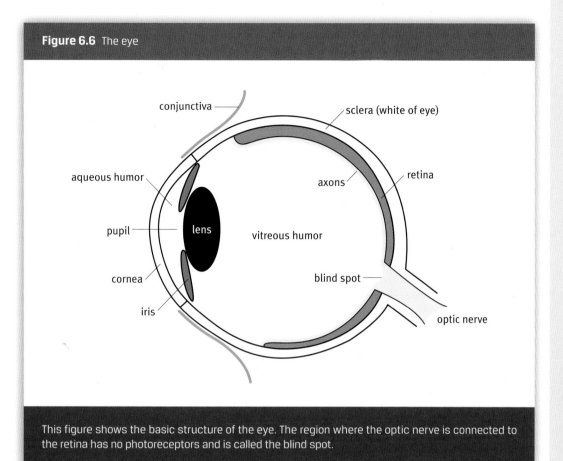

This figure shows the basic structure of the eye. The region where the optic nerve is connected to the retina has no photoreceptors and is called the blind spot.

The process of photoreception

Light striking rods and cones is captured by photopigment molecules called opsins, initiating the process of transduction, whereby light is converted into a synaptic signal. The end-point of this process is closure of sodium channels in the receptor membrane. This is the opposite of what happens in other sensory systems, where signals are generated by the opening of sodium channels. Closure of sodium channels in photoreceptors causes hyperpolarization of the membrane and a reduction in the amount of neurotransmitter released by photoreceptor axons. Bipolar cells respond to these neurotransmitter interruptions in one of two ways. Half of the bipolar cells hyperpolarize like the photoreceptors, causing a decrease in their release of glutamate. The others invert the signal from photoreceptors so that they depolarize and increase their transmitter release. These are referred to as the OFF- and ON- responses, respectively. Signals from the bipolar cells are received by the ganglion cells, which generate action potentials. These action potentials travel along the optic nerve to send visual information to the brain.

The projection to the brain

A vertical line through the macula represents the division between the temporal retina (lateral side) and the nasal retina (medial side). Axons from ganglion cells in the nasal retina cross the midline in the optic chiasm so that their information goes to the visual centers of the opposite side of the brain. The axons from the temporal retina do not cross the midline and their information goes to

Cataract–a common cause of blindness

A serious, but entirely curable, form of blindness is caused by the formation of a cataract in the lens. The cataract is an expanding area of opacity that in the end will block all useful vision in that eye. In cataract surgery, the affected lens is removed and replaced with a plastic lens. The result is an immediate return to clear vision.

Figure 6.7 The layers of the retina

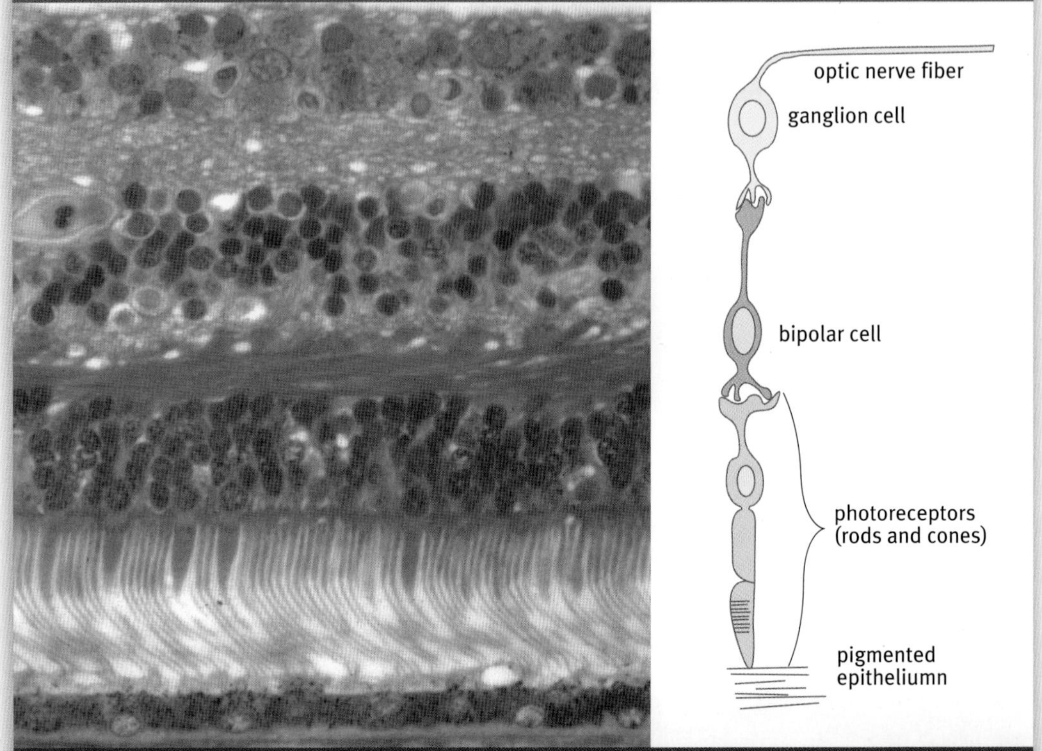

Labels on diagram: optic nerve fiber; ganglion cell; bipolar cell; photoreceptors (rods and cones); pigmented epitheliumn

The microstructure of the retina is shown in a photomicrograph on the left of this figure. Light passes through all the superficial layers of the retina (optic nerve fiber layer, ganglion cell layer, bipolar cell layer, and photo receptor cell layer) before it reaches the rods and cones, which are sensitive to light. The photoreceptors lie on the surface of the pigmented epithelium of the choroid layer of the eye. Outside the choroid is the hard scleral layer of the eye. The diagram at the right of the photomicrograph shows the neuronal pathway from a photoreceptor cell to a bipolar cell, and then from a ganglion cell to the optic nerve.

Damage to the optic chiasm

The optic chiasm lies just in front of the pituitary stalk, so tumors of the pituitary can press against the chiasm and cause loss of vision. Because the crossing optic nerve fibers come from the nasal retina, the lateral (temporal) visual field is affected. If the chiasm is completely damaged, the person will only have vision immediately in front of them, and the lateral parts of the visual field will appear black.

---

the cerebral cortex of the same side. The situation is basically the same in rodents, except that the area of temporal retina is relatively small–about 10% of the total area of the retina.

To understand the way the visual world is transmitted to the cortex, we have to consider the different paths taken by axons from each half of the retina (Figure 6.8). The axons from the medial half (the nasal retina) all cross in the chiasm, whereas those from the lateral (temporal) half do not cross. Because light reaches the retina through a lens, the light received from the temporal half of the visual field falls upon the nasal parts of each retina, and vice versa. Fibers from the nasal and temporal retina travel together in the optic nerve, but are separated at the chiasm. The image from each temporal visual field (nasal retina) crosses in the chiasm to continue as an optic tract on the opposite side of the brain. The image from each nasal visual field (temporal retina) remains uncrossed and continues on to reach the thalamus and cortex of the same side as the retina. The outcome is that the left-hand sides of the visual fields of both eyes run in the right optic tract, and the right-hand sides of the visual fields of both eyes run together in the left optic tract. The overall result is that each visual cortex is looking at the visual world of the opposite side. In other words, the right retina of both eyes projects to the right hemisphere, and vice versa.

It must be appreciated that what we have described here is the pattern in humans. In most mammals, over 90% of the optic nerve fibers cross over in the optic chiasm, and only about 10% remain uncrossed. In animals with eyes pointing laterally, such as rats, this means that the visual world of one side reaches the opposite cortex in an uncomplicated way. The reason the human system is more complicated is that our eyes are placed at the front of our head and the two eyes have overlapping visual fields. The resulting binocular vision in humans gives us a high degree of stereoscopic vision compared to other mammals.

The visual information humans are most interested in is very precise (like the fine detail on a map). However, there are a number of separate systems for different functions that are related to light reception. For example, our internal 24-hour clock (a nucleus situated in the hypothalamus just above the optic chiasm) is interested in light and dark periods for entraining (setting the clock). Even more elaborate visual subsystems are found in birds. Eagles and hawks have one system for detecting prey from the air, one for scanning the horizon, and another for looking at food at the end of the beak. Birds even have a special system for collision avoidance, similar to that used in modern commercial airliners.

Hearing

In much the same way that the optics of the eye affect visual information before it is detected, the human ear shapes and transforms sound waves before they are detected by the cochlea. The folds of cartilage surrounding the ear canal are called the pinna, and serve to reflect and attenuate sound.

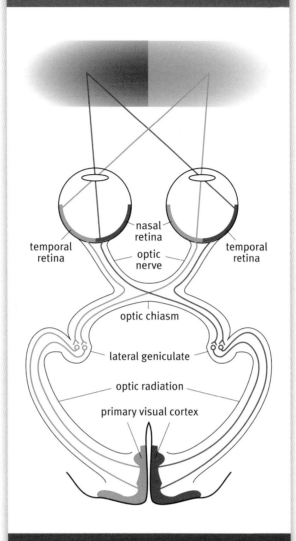

Figure 6.8 Human visual pathways

The right and left halves of each visual field are reversed in their projection onto the retina so that the temporal (lateral) retina looks at the nasal (medial) half of the visual field. The nasal retina looks at the temporal half of the visual field. The fibers that arise in the nasal half of the retina cross in the optic chiasm, and the information they convey reaches the opposite visual cortex. The fibers that arise in the temporal half of the retina do not cross the midline, so that their information is sent to the visual cortex of the same side. The net effect of this arrangement is that each visual cortex receives images from the opposite visual fields. In this diagram the left visual fields are colored red and the retina that receives the information is also colored red. The pathway from the red retina can be followed through the thalamus to the visual cortex. The pathway from the right visual fields to the cortex is shown in blue. (Adapted from Brodal, 2004, p 194)

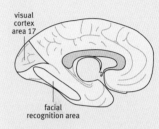

Figure 6.9
Visual cortex and the facial recognition area

Visual information is primarily received by the primary visual area of the cortex (area 17, as known as V1). Area 17 is located in a wedge shaped area on the medial side of the occipital pole. Close to area 17 are a number of secondary visual areas, which have specialized roles in feature extraction. An outstanding example of this is the facial recognition area which is located along the fusiform gyrus on the medial side of the temporal lobe. The midbrain and hindbrain have been removed for clarity.

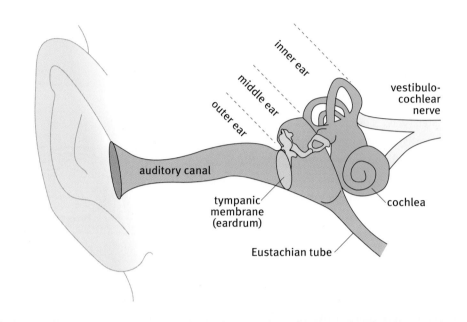

Figure 6.10 Components of the ear

Sound enters the external ear through the auditory canal and causes the tympanic membrane to vibrate. This vibration is transmitted across the middle ear by three tiny bones, the malleus, incus and stapes. The stapes transmits the vibration to fluid in the internal ear, and the vibrations eventually reach the cochlea. The cochlea is a bony shell which contains the sound receptors. The middle ear is connected to the back of the throat by an air filled canal called the Eustachian tube. (Adapted from Campbell, 1996, p 1038)

Deafness

Deafness can be caused by damage or blockage to any part of the ear. In the external ear, accumulation of wax can cause deafness. Infections of the middle ear can cause deafness by impeding the movement of the auditory ossicles. In the internal ear, the most common cause of partial or complete deafness is the destruction of hair cells of the organ of Corti due to very loud noise. Tumors on the auditory (acoustic) nerve can also cause deafness.

As we turn our heads, this changes the sound in ways which provide additional information to the brain, and aid in determining the direction from which the sound came. The auditory canal helps to amplify the sound. At the end of the canal lies the tympanic membrane, which marks the beginning of the middle ear.

The middle ear contains three tiny bones—the malleus, the incus, and the stapes. When sound waves enter the middle ear these bones act as levers and convert the lower-pressure eardrum sound vibrations into higher-pressure sound vibrations in the fluid of the inner ear.

The inner ear consists of the cochlea and balance receptors. The cochlea is the auditory receptor organ. It has three fluid-filled chambers: the scala media, scala vestibuli, and scala tympani. The latter two contain perilymph whereas the scala media contains endolymph. The chemical difference between the two fluids is important for the function of the inner ear. The ion concentrations in perilymph are similar to those in extracellular fluid, whereas the ion concentrations in endolymph are similar to those in intracellular fluid. The three scalae are twisted into a spiral ('scala' is Latin for 'staircase') and these three form three spiral staircases.

Lying in the floor of the scala media is a strip of sensory epithelium called the organ of Corti. The principal cells of the organ of Corti are the hair cells, which transform fluid waves into electrical nerve impulses. The hair cells are columnar and covered in bundles of specialized cilia. Resting

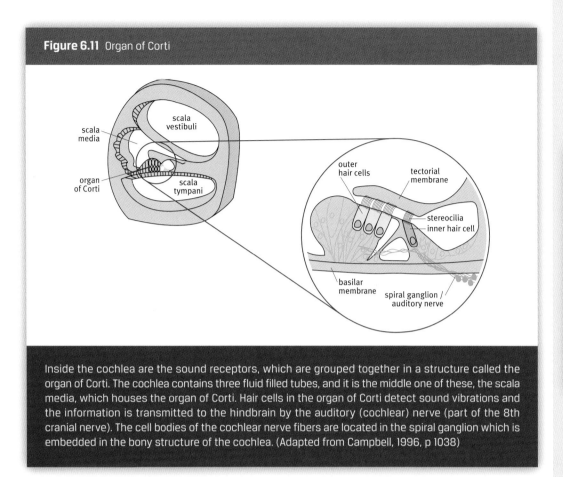

Figure 6.11 Organ of Corti

Inside the cochlea are the sound receptors, which are grouped together in a structure called the organ of Corti. The cochlea contains three fluid filled tubes, and it is the middle one of these, the scala media, which houses the organ of Corti. Hair cells in the organ of Corti detect sound vibrations and the information is transmitted to the hindbrain by the auditory (cochlear) nerve (part of the 8th cranial nerve). The cell bodies of the cochlear nerve fibers are located in the spiral ganglion which is embedded in the bony structure of the cochlea. (Adapted from Campbell, 1996, p 1038)

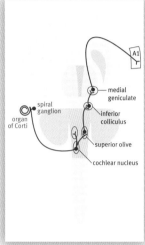

Figure 6.12
The pathway from the organ of Corti to the cerebral cortex

The axons of the spiral ganglion cells reach a group of cochlear nuclei in the hindbrain. The cochlear nuclei project to another neuronal cluster called the superior olive. The projection to the superior olive is both crossed and uncrossed. Most fibers leaving the superior olive do not project directly to the thalamus instead they synapse in the inferior colliculus which then connects with the auditory thalamus (medial geniculate nucleus). The medial geniculate projects to the auditory cortex. (Flat map adapted from Swanson, 2004)

on the surface of the cilia is the tectorial membrane, which moves back and forth with each cycle of sound, tilting the cilia and changing the membrane potential of the hair cell.

The inner hair cells are connected to the neurons of the auditory nerve. The detected stimuli travel along the auditory nerve to the cochlear nuclei and the superior olivary complex of the hindbrain. From here, the information is sent to the inferior colliculus of the midbrain, where it is processed before being sent to the medial geniculate nucleus in the thalamus. The medial geniculate projects to the auditory cortex in the temporal lobe.

Vestibular system

The sensory system which provides information on balance and head position is called the vestibular system. The name comes from the vestibule, which is a small bony space in the inner ear in which the receptors are located. The vestibule is next to the cochlea, and the two structures together are housed in the bony labyrinth of the inner ear.

The vestibule contains two quite separate systems for gathering information on head postion. One system, the semicircular canals, detects movement of the head in any direction, while the other system, the saccule and utricle system, can register the position of the head when it is not moving. The first system is described as the dynamic vestibular system, and the second is referred to as the static vestibular system. Like the cochlea, these two systems rely on information gathered by sensitive hair cells. There are three patches of hair cells in the semicircular canals and one each for the saccule and utricle.

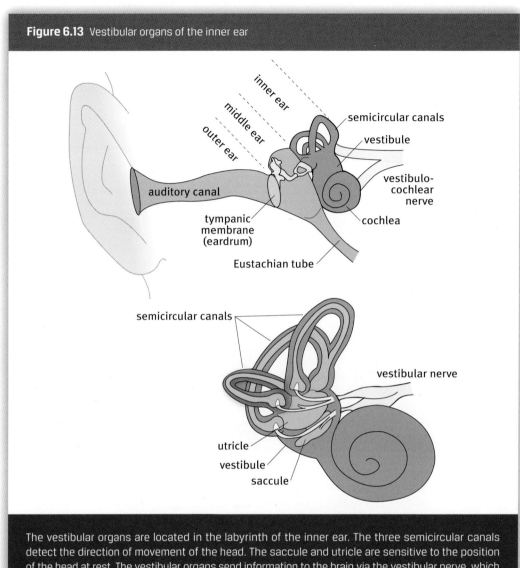

Figure 6.13 Vestibular organs of the inner ear

The vestibular organs are located in the labyrinth of the inner ear. The three semicircular canals detect the direction of movement of the head. The saccule and utricle are sensitive to the position of the head at rest. The vestibular organs send information to the brain via the vestibular nerve, which joins the cochlear nerve to become the eighth cranial nerve, the vestibulocochlear nerve. (Adapted from Campbell, 1996, p 1040)

The main function of the vestibular system is to maintain balance and to coordinate eye movement with head and body position. These functions are managed at a reflex level, with constant adjustments being made to postural muscles and eye movement. Input from the vestibular system enables the eyes to track an object of interest, no matter how the body is moving. The coordination of eye movement uses a network including the vestibular system, the visual system, the cerebellum, the spinal cord, and motor neurons of the eye muscles.

As well as controlling this complex reflex network, the vestibular system supplies information to the cerebral cortex about the position of the body in space. This helps to give us a conscious appreciation of the stability, or otherwise, of the body. However, the vestibular system is not the only source of information on body position in space. The visual system and the somatosensory system both provide vital information that contributes to our ability to maintain posture and balance. Some of the position sense information from the somatosensory system comes from touch receptors in the skin and deep pressure receptors in the feet.

Figure 6.14 Semicircular canals and the cupula

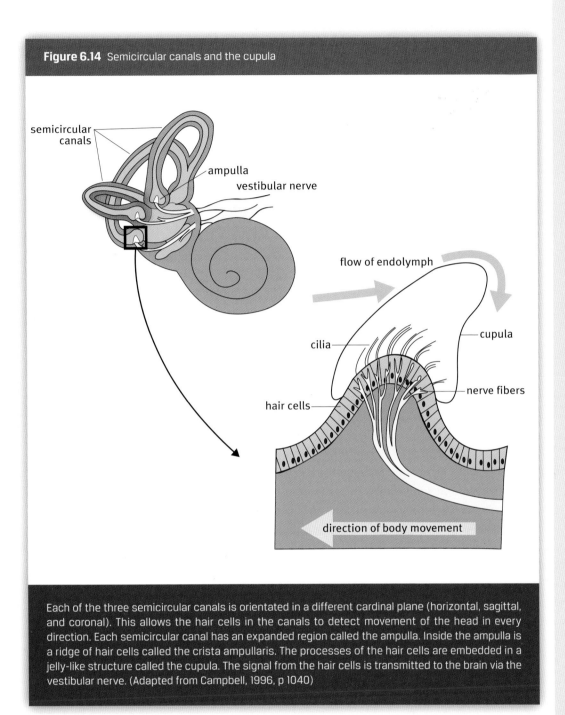

Each of the three semicircular canals is orientated in a different cardinal plane (horizontal, sagittal, and coronal). This allows the hair cells in the canals to detect movement of the head in every direction. Each semicircular canal has an expanded region called the ampulla. Inside the ampulla is a ridge of hair cells called the crista ampullaris. The processes of the hair cells are embedded in a jelly-like structure called the cupula. The signal from the hair cells is transmitted to the brain via the vestibular nerve. (Adapted from Campbell, 1996, p 1040)

Semicircular canals

The semicircular canals are three almost circular tubes, which contain fluid called endolymph. Endolymph is also found in the cochlear duct in contact with auditory hair cells. Each canal is placed in a different spatial plane–one horizontal, one vertical close to the sagittal plane, and one vertical close to the coronal plane. No matter what direction the head moves in, it will cause fluid to move around in one or more of the three canals. Each canal has a widened area called the ampulla, which contains the sensitive hair cells. The processes of the hair cells are embedded in a piece of jelly called the cupula. When the fluid moves in response to head movement, the cupula is dragged in the direction of fluid movement, twisting the processes of the hair cells. The twisting generates an electrical signal, which is carried to the hindbrain by the vestibular nerve.

Saccule and utricle

The saccule-utricle system is contained in a pair of recesses in the vestibule. The saccule and utricle both have a patch of hair cells called a macula. The processes of these hair cells are embedded in a plate of jelly, but in this case, the jelly contains many calcium carbonate crystals called otoliths. The calcium-containing jelly plate is called the otolith membrane. When the head is stationary, the weight of the otoliths pulls the hair cells under the influence of gravity, and this in turn causes the hair cells to be stimulated. Other external forces such as rapid acceleration can also stimulate the hair cells in the macula. The receptor clusters of the saccule and utricle are set at right angles to each other, and can therefore detect changes in the position of the head at all times. The macule of the saccule is set in the transverse plane, so it is best placed to detect flexion and extension movements of the head. The macule of the utricle is set in the sagittal plane, so it is best placed to detect lateral tilting of the head. The information gathered by the receptors of the saccule and utricle is carried to the hindbrain by the vestibular nerve. The ganglion cells of the vestibular nerve fibers are true bipolar cells. The vestibular ganglion is located next to the vestibule.

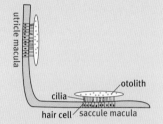

Figure 6.15
Utricle and saccule

This is a diagrammatic representation of the saccule macula and utricle macula. Each of these two structures is made up of a patch of hair cells whose processes are embedded in a jelly-like plate. Embedded in the jelly are many small calcium carbonate crystals called otoliths. The maculae of the saccule and utricle are positioned at right angles to each other for optimal detection of head position. The signal generated by the hair cells is transmitted to the brain via the vestibular nerve.

Vestibular nuclei

The vestibular nuclei in the hindbrain occupy a large area in the floor of the fourth ventricle. They are strongly connected with the cerebellum, the spinal cord, and with the system that controls eye movement. The relationship between the vestibular receptors and eye movement can be easily demonstrated. If you sit someone in a rotating office chair and spin them around ten times at a moderate speed, then stop them abruptly, you will see the eyes flick from side to side in a movement called nystagmus. Because the fluid in the horizontal semicircular canal is still moving when chair stops moving, the eye muscle control system thinks that the head is still moving, and responds with side-to-side movement of the eyes. The subject will also complain that the room seems to be spinning around them; this sensation is called vertigo. It is important to distinguish the sensation of vertigo from dizziness. Dizziness is a feeling of unsteadiness, while in vertigo the world seems to be moving around you. Vertigo is often accompanied by feelings of nausea, and may cause vomiting.

The cerebellum and the vestibular system

The cerebellum is responsible for the coordination of voluntary movement. In humans and other primates, it plays a particularly important role in the coordination of eyes and hands. Vestibular input to the cerebellum enables the cerebellum to correct every movement for changes in head position and rotation of the trunk. When this system is compromised, such as in alcohol intoxication, walking and reaching movements become increasingly disturbed.

Vestibulospinal tracts

The vestibular nuclei give rise to two fiber tracts that connect with the spinal cord. The medial vestibulospinal tract is controlled by input from the semicircular canals and connects with the neck muscles to control head position. The lateral vestibulospinal tract is controlled by the saccule and utricle and connects to the whole spinal cord. Its main role is to adjust posture in the limbs and trunk by activating extensor muscles.

Vestibular representation in the cerebral cortex

While the main function of the vestibular system is the automatic control of body and eye position, the cerebral cortex receives vestibular information and this contributes to a conscious perception of body position. The main area receiving vestibular information in monkeys is located between the ventral part of the primary somatosensory cortex and the insula; it is assumed that an equivalent area exists in humans. Because the primary somatosensory cortex is in the post-central gyrus of the parietal lobe, this area is referred to as the parietal-insular vestibular cortex (PIVC). The input to this area is not exclusively vestibular; it also receives visual input and information from position-sense receptors in the neck. There are a number of other small areas in the parietal lobe that have been shown to have vestibular input (such as area 2Ve), but it seems that the PIVC has a central role in combining all available information on body position to provide the cortex with an accurate sense of the position of the body.

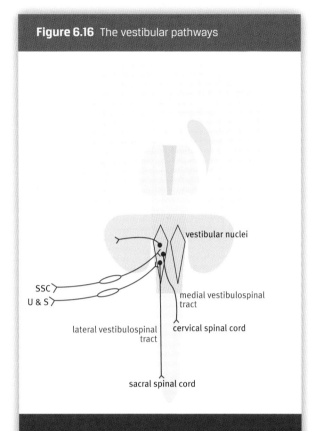

Figure 6.16 The vestibular pathways

The primary sensory neurons in the vestibular system are bipolar neurons. The peripheral axon of each bipolar cell is connected to the semicircular canals (SCC) or to the saccule (S) or utricle (U). The central axon of each bipolar cell is connected to the vestibular nuclei in the floor of the fourth ventricle. The semicircular canal bipolar cells connect with the medial vestibulospinal tract which controls head position. The saccule and utricle bipolar cells connect with the lateral vestibulospinal tract which controls the position of the trunk and limbs. The vestibular nuclei are tightly connected with the cerebellum. (Flat map adapted from Swanson, 2004).

Taste sensation

The basic sense of taste is relatively crude, because the taste receptors can only distinguish a limited number of characteristics of food. We therefore rely on our sense of smell to judge the quality of what we eat. As we chew or swallow, the odors of the food reach the olfactory nerve endings and supplement the information from the taste receptors on the tongue.

Taste

Taste receptors in the tongue send information via the 7th and 9th cranial nerves to reach the nucleus of the solitary tract in the hindbrain. The nucleus of the solitary tract projects to a specific gustatory nucleus in the thalamus, and from there to the insular cortex. The taste receptors in the tongue have only a limited range of perception (salt, sweet, sour, bitter, savory, and piquant) and much of what we call taste is actually based on extra information from our olfactory system.

Smell

In most mammals, smell (olfaction) plays a critical role in the recognition of predators, partners, and offspring. In humans, the olfactory system is less important because of our reliance on a highly developed visual system. However, it is not often realized that a great deal of what we call taste is based on olfactory signals. When we have a nasal infection and the olfactory receptors are injured or blocked, we lose about 90% of our ability to appreciate flavors. In rodents, olfactory input to the

Pheromones

Pheromones are signaling chemicals released into the environment to cause behavioral changes in nearby individuals. Insects use pheromones to signal a wide range of situations, including sexual readiness, alarm, and to mark a trail. Most mammals have a small pheromone detecting system called the accessory olfactory system. It is primarily related to sexual signaling. This system is not present in humans.

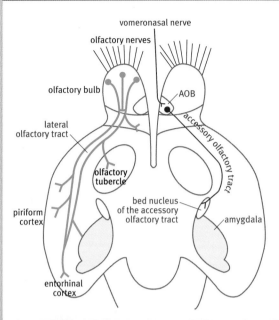

Figure 6.17 The olfactory pathways

The main olfactory pathway runs from the olfactory bulb to the primary olfactory cortex (piriform cortex) on the underside of the temporal lobe. In almost all animals, but not in humans, there is an accessory olfactory pathway that is exclusively sensitive to pheromones. Pheromones are sexual attractants that trigger sexual behavior. The accessory olfactory pathway comprises the vomeranasal nerve (not present in humans), the accessory olfactory bulb, the accessory olfactory tract, and a small nucleus in the amygdala. (Adapted from Nieuwenhuys et al, 1997, p 1896)

amygdala can initiate emotional responses, including aggression.

The olfactory epithelium of the nasal cavity is the site that converts odorant information into coded neural signals. Olfactory receptor neurons in the roof of the nose are subject to continual damage, and in rodents they have a lifespan of only 40 days. Because of this, they are constantly replenished from stem cells in the area. An unmyelinated axon arises from each olfactory receptor neuron. Bundles of these axons travel through holes in the ethmoid bone in the roof of the nose to reach the olfactory bulb.

In addition to the main olfactory system, many mammals (but not humans) have a second olfactory system concerned with the detection of pheromones. Pheromones are odorants that trigger behavioral responses, such as the desire to mate. The pheromone-sensitive area is called the vomeronasal organ. Removal of the vomeronasal organ in juvenile male mice does not impair food-finding abilities, but does impair their ability to recognize sexual signals from potential mates. Axons of receptor neurons in the vomeronasal organ project exclusively to the accessory olfactory bulb (AOB), which lies adjacent to the main olfactory bulb. The output of the accessory olfactory bulb is to a special nucleus superficial to the amygdala, called the bed nucleus of the accessory olfactory tract.

The olfactory bulb has a complex internal circuitry for processing of olfactory information. It sends fibers to the piriform cortex and other primary olfactory cortical areas. The fibers projecting from the olfactory bulb mainly travel in the lateral olfactory tract and in the stem of the anterior commissure. The anterior commissure pathway connects one olfactory bulb to the one of the opposite side. In humans, the primary olfactory cortex is located in a region called the uncus at the rostral tip of the temporal lobe.

The olfactory ventricle is a small space filled with cerebrospinal fluid, in the middle of the olfactory bulb. It is formed from an extension of the lateral ventricle. In rodents, an area directly under the lateral ventricle, the subventricular zone (SVZ), produces stem cells that migrate forward from the lateral ventricle into the olfactory bulb. They provide a continual source of new neurons for the olfactory system throughout adult life. This cell group is called the rostral migratory stream.

Sensory processing outside the cortex

At the start of this chapter we mentioned that conscious perception is just one aspect of sensory function, since the information collected by sensory receptors is registered at many levels in the nervous system. Every sensory axon entering the nervous system gives off collaterals, contacting many other neurons and systems along the way. The cerebral cortex is an important destination for sensory information, but not in every case; in fact, the cortex literally receives sensory information on a 'need to know' basis. Many sensory events trigger neural responses without ever reaching the cortex. For example, our blood pressure is continuously monitored by baroreceptors, stimulating receptor neurons whose axons trigger reflexes to adjust heart rate and force, without us ever being aware of our own blood pressure. The nervous system of the gut (enteric nervous system, see Chapter 4) has a range of sensors which detect pH and the composition and consistency of gut contents. This information is used to adjust bowel contraction and trigger secretions as needed, while the conscious mind remains completely unaware of digestion after food is swallowed. However, we are aware of sensory events which are directly relevant to planning and choosing behavior.

An example: rolling an ankle

Imagine rolling your ankle while you run, causing an unexpected stretch to the muscles on the lateral side of the calf. The stretch receptors in these muscles fire rapidly, and the information is sent to the spinal cord via the dorsal roots of a lumbar spinal nerve. From here, each axon makes synaptic connections with hundreds of reflex circuits in the spinal cord, sending collaterals to pattern generators for walking, running and jumping. Finally, the main stem of the axon will travel to the hindbrain to synapse in the dorsal column nuclei. From there the second order neuron may send collaterals to brainstem pattern generators, to the cerebellum via the inferior olive, and to midbrain locomotor centers, on its way to synapsing with thalamic neurons, that stimulate the somatosensory cortex. The obvious message here is that one signal triggers many synaptic events.

The input from the stretch receptors in this one muscle group will help stimulate dozens of reflex adjustments in the spinal cord, helping to activate muscles that stabilize the ankle. It will also be used to adjust tension in the muscle itself, and moderate the activity of other muscles which act on the ankle. The stretch information will be used to adjust the spinal cord pattern generator regulating running movements, by breaking the stride and allowing time to rebalance. Trunk muscles may be activated to shift the center of gravity of the body to help maintain balance. Brainstem pattern generators guiding the running may change their activity, slowing down in case of injury or overbalancing. In the cerebellum, information about balance, posture and the position of the body in space will be updated in order to adjust subsequent movements. If the runner is in danger of falling, the cerebellum will trigger powerful balance reflexes to correct posture and prevent a fall.

Most of these reflexes and processing will be triggered before the information reaches the cortex. By the time the runner becomes aware of having rolled the ankle, the rest of the nervous system will have made the necessary adjustments, at the same time as dealing with many other sensory events. Most of those adjustments will be completely unconscious–handled by nervous system components outside the cortex.

The value of conscious perception becomes clear, however, if the ankle roll is severe enough to cause the runner to fall. Such an event is outside the range of problems that can be handled by

---

Convergence and divergence

In the nervous system, the concept of 'convergence' refers to multiple axons synapsing on a common target, such as the alpha motor neurons described in Chapter 5. 'Divergence' refers to the activity of a single axon which is distributed to multiple targets, such as the ankle stretch which triggers multiple systems. Most information in the nervous system is a combination of convergent and divergent activity, allowing stimuli to be compared and processed.

pattern generators and simple reflexes, and requires a more strategic response. A serious injury might require rest, and this may also require a strategy to reach somewhere safe to rest. Cortical processing is essential to plan remedial actions, like choosing safer ground to walk on, and deciding how much pain to tolerate in order to reach a resting place.

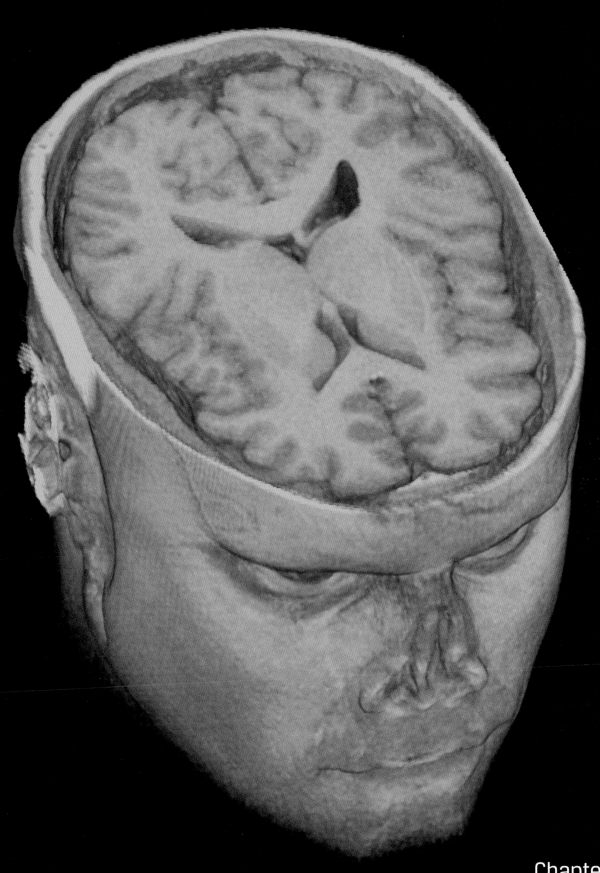

Chapter 7

The human cerebral cortex

7 The human cerebral cortex

In this chapter:

The cerebral cortex–anatomy and histology 98

Guiding principles of cortical structure and function 102

The functional layout of the human cerebral cortex 103

The cerebral cortex and behavior 106

The functions of the brain are intimately related to our behavior and personality–the qualities we most readily identify as human. Our limbs or organs can be damaged, removed or transplanted without affecting our 'human' qualities, but even minor damage to the brain can radically change consciousness, perception, behavior, or personality. This naturally prompts us to ask–what is special about the human nervous system?

Other animals lack the behavioral sophistication and abstract intelligence of human beings, and since these seem to be functions of the nervous system, investigators have studied the differences between humans and other mammals, seeking clues to the basis of human thought and intellect. In this book, we have used the rat brain as a guide to the organization of the human brain, since the brains of humans and other animals are far more similar than they are different. As far as we know, human neurons and glia function near-identically to those of other mammals, and the physiology of human nervous systems is similar to that of other species. The overwhelming majority of genes expressed in the human nervous system are identical, or closely similar, to those of other mammals, to the extent that human genes can be inserted into transgenic mice and function similarly. Since nervous systems evolve by incremental change, adapting and reusing available mechanisms for new purposes, it is unsurprising that there are strong genetic similarities between animals with shared ancestry. Instead, the special characteristics of the human brain lie in how we configure and use fairly simple functional units to achieve human consciousness and behavior. In particular, our complex behaviors rely on the cerebral cortex.

The mammalian cerebral cortex has several unique qualities that have allowed it to develop rapidly in a short space of evolutionary time, without disrupting the rest of the nervous system. In particular, its role is to analyze the senses and direct the activity of the rest of the nervous system, without being involved in basic housekeeping and bodily regulation. The cortex derives a sophisticated understanding from sensory inputs, combining them with memory and experience to produce behavior that is more useful than simple reflexes and reactions. This overseer role of the cerebral cortex allows it to grow and reconfigure its abilities, without compromising the core survival functions of the nervous system. This is clearly demonstrated in recovery from a cerebral stroke, in which the basic life-sustaining abilities of the brain are maintained, even though the cortex may be severely damaged.

The analytical abilities of the cortex are the product of its cellular architecture and the way it communicates with the rest of the nervous system. This chapter will deal with the cellular and anatomical properties of the cortex, as well as its functional abilities.

The cerebral cortex–anatomy and histology

The human nervous system shares its basic layout and functional arrangement with other mammals. However, the human cerebral cortex is huge and complicated compared with that of rodents, forming an overgrowth that bulges out and distorts the shape of structures lying inside the hemispheres, including the ventricles, the striatum, and the hippocampal complex. The human cerebrum expands forwards to form the frontal lobe, backwards to form the occipital lobe, and grows downwards and outwards into the temporal lobe. These expansions hide part of the cortex that does not grow as fast—a buried area called the insula, concerned with taste and gut control, which is found on the surface of the cortex in rodents.

The word cortex means an outer layer, such as the bark of a tree or the skin of an orange; the cerebral cortex is a sheet of neurons and glia covering the rest of the forebrain. In rodents it is a flat sheet, but the human cortex expands so much that it is thrown into folds (called gyri) and grooves (called sulci). Beneath the cortex is the white matter, a dense mat of axon fibers joining cortical regions and communicating with the rest of the nervous system.

Uniformity of structure

One of the most striking features of the cerebral cortex is how little its cellular structure varies from region to region, even between areas dealing with very different types of sensory input. This reflects the abstract nature of cortical processing, which is not so much concerned with the specifics of receptors and effectors, as with the meaning of sensory stimuli and their relevance to behavior. Most of the human cerebrum consists of neocortex, the part that has most recently evolved in mammals. The neocortex is also called isocortex, because a standard set of cellular arrangements is repeated across the entire sheet ('iso' means uniform or unchanging). As a consequence, neuroanatomists trying to distinguish one cortical area from another have to analyze the thickness of cellular layers and the sizes of cells found there. Even then the boundaries in many cases are not obvious, because each area merges with its neighbors. The clearest distinction is between granular (sensory) cortex and agranular (motor) cortex. In the sensory cortical areas, the spiny stellate cells form a dense granular layer, whereas the motor cortex is agranular because it lacks these cells.

Cortical neurons

The circuits of the cortex are made up of a variety of neuronal types, each with characteristic neurotransmitters, chemistry and connections. Most of them are excitatory (using glutamate or aspartate as a neurotransmitter), including the most common type, the pyramidal cell (Figure 7.2). These cells typically have a conical cell body with one large dendrite at the apex reaching toward the cortical surface, and other dendrites spreading out from the base of the cone. The dendrites are covered in small receptive structures called spines, a property they share with the other main

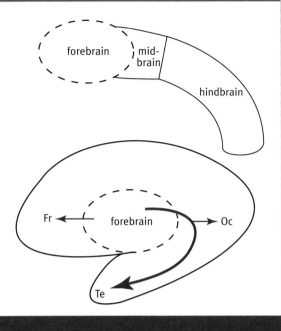

Figure 7.1 Expansion of the lateral part of the human forebrain during development

This is a diagram of the enlargement of the lateral parts of the forebrain during human CNS development. During this time, the lateral (telencephalic) part of the forebrain expands dramatically to form the large frontal (Fr), occipital (Oc), and temporal (Te) lobes. The overall shape created by these expansions is in the form of a letter 'C.' Many forebrain structures are forced to assume this 'C'-shape. The 'C'-shaped structures include the fornix, the lateral ventricle, and the caudate nucleus.

Dendritic spines

Pyramidal and spiny stellate cells of the cortex have dendrites bristling with short narrow extensions called spines. The tip of each spine bears the post-synaptic receptors for an excitatory synapse with a passing axon in the neuropil. Spines are the basis for most of the connectivity and plasticity of these neurons; they can extend to form new connections, or retract to disconnect, in a matter of hours.

excitatory cell, the spiny stellate cell. Pyramidal cell axons travel from the base of the cone toward the white matter, where they may branch out to nearby cortex, to the cortex of the opposite side, or to structures outside the cerebrum, like the thalamus, brainstem, and spinal cord. Spiny stellate cells receive input from the thalamus, and have axons that excite cells in layers above and below them.

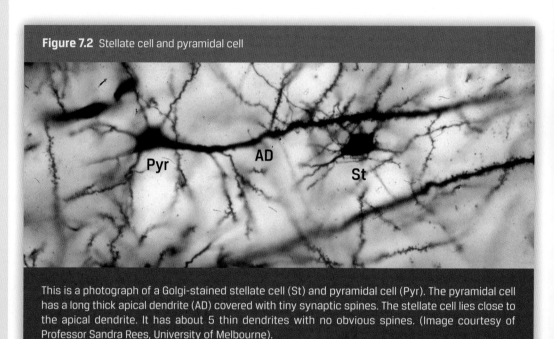

Figure 7.2 Stellate cell and pyramidal cell

This is a photograph of a Golgi-stained stellate cell (St) and pyramidal cell (Pyr). The pyramidal cell has a long thick apical dendrite (AD) covered with tiny synaptic spines. The stellate cell lies close to the apical dendrite. It has about 5 thin dendrites with no obvious spines. (Image courtesy of Professor Sandra Rees, University of Melbourne).

Golgi and Cajal

In the 1880s, the Italian scientist Camillo Golgi developed the amazing reduced silver stain that bears his name. The stain launched a revolution in neuroanatomy, because it allowed scientists for the first time to see the complete anatomy of dendrites and axons of neurons. The Spanish histologist Santiago Ramón y Cajal used this technique to make meticulous pen-and-ink drawings of neurons in all parts of the brain. Cajal and Golgi shared the Nobel Prize in 1906.

Inhibitory cells make up about 20% of cortical neurons, and nearly all of them use GABA as a neurotransmitter. Their function varies according to the arrangement of their axons and dendrites, which enable specific types of inhibition. For instance, basket cells and clutch cells send axon terminals to the bodies of pyramidal cells and spiny stellate cells, whereas the chandelier cell gives rise to processes that wrap around the initial segments of pyramidal cell axons, in effect governing whether they are able to fire or not.

Cortical layers

Under the microscope, the neurons of the cerebral cortex can be seen to be arranged in distinct layers. In evolutionarily older areas of cortex, such as the olfactory cortex, there are only three layers, but in the neocortex, six layers are traditionally recognized. The numbering of the layers is from the surface to the deep layers, and they are numbered from one to six. The earliest attempts to map the human cerebral cortex were based almost entirely on examination of subtle differences in the layering pattern from one cortical area to another. Brodmann and other pioneers of the early 20th century had to rely on Nissl-stained sections and myelin-stained sections to delineate the cortical areas. This work was very difficult and very tedious, but it enabled comparison of human cortical areas with those in monkeys. Fortunately, the major motor and sensory areas can be recognized in non-human animals on the basis of distinctive layering patterns. The recognition of comparable areas in non-human mammals allowed scientists to gather data on connections and functions of particular cortical areas. This information could then be used to help understand the functional organization of the human cerebral cortex.

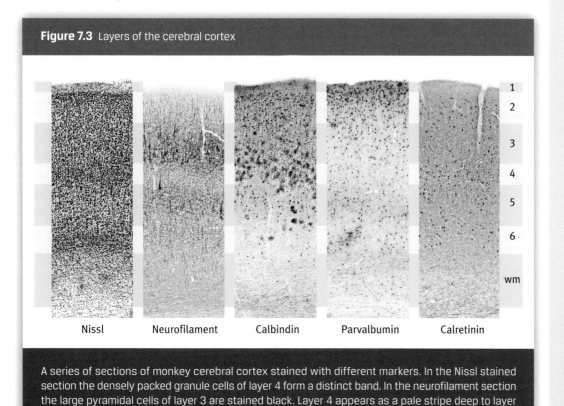

Figure 7.3 Layers of the cerebral cortex

A series of sections of monkey cerebral cortex stained with different markers. In the Nissl stained section the densely packed granule cells of layer 4 form a distinct band. In the neurofilament section the large pyramidal cells of layer 3 are stained black. Layer 4 appears as a pale stripe deep to layer 3. In the other three markers, some of the layers are differentialy stained (wm=white matter). (Images courtesy of Dr Hironobu Tokuno. Tokyo Metropolitan Institute for Neuroscience)

Each layer in the isocortex has a characteristic set of cells and connections.

Layer 1 is a dense mat of fibers, which spread back and outward across the surface of the cortex. Many of them carry specialized transmitters, such as noradrenaline and dopamine from the hindbrain, penetrating deeper layers to distribute them throughout the cortex.

Layer 2 consists of small pyramidal cells and small inhibitory cells, which act in local circuits of neurons to process inputs.

Layer 3 has larger and more numerous pyramidal cells, which send excitatory outputs to other parts of the cortex, either nearby or in the opposite hemisphere. Basket cells and other inhibitory cells in layer 3 spread inhibition to neighboring cells, as part of cortical processing.

Layer 4 is the main recipient of sensory fibers coming from the thalamus. Its granular appearance is due to the concentration of small stellate cells. In specialized sensory areas, the spiny stellate cells of layer 4 are organized into a number of sub-layers. In motor regions layer 4 granule cells are absent, but a ghost of the layer still remains.

Layer 5 contains the largest pyramidal cells, which send their axons to regions outside the cortex, such as the striatum and spinal cord. The largest layer 5 cells, called Betz cells, are found in the primary motor cortex and send their axons to form the corticospinal tract.

Layer 6 has small pyramidal cells that project to part of the thalamus, which in turn supplies inputs to the cortex. This arrangement completes a feedback loop which regulates incoming thalamic activity.

Arabic or Roman numerals?

It has been fashionable for over a century to use Roman numerals to label cortical layers, cranial nerves, and subdivisions of other structures in the nervous system. Unfortunately, the one (I) and five (V) in the Roman system can be confused with capital I and capital V in the alphabet. Use of Arabic numbers removes this cause of confusion, with no loss of accuracy. We strongly recommend that you use Arabic numerals instead of Roman numerals wherever possible.

Guiding principles of cortical structure and function

Although the cerebral cortex is undeniably complex, some useful generalizations can be made about how its structure is related to its functions.

A balance between excitatory and inhibitory activity

If excitatory activity in the cortex were not tightly controlled, storms of cortical activity would result. Such storms can cause epileptic attacks. Excitatory activity is controlled by inhibitory interneurons in each of the cortical areas. This results in a balance between necessary excitation and the risk of uncontrolled spread of electrical activity. Neurotransmitters such as dopamine, serotonin, and noradrenaline tend to have modulatory or enhancing effects within the cortex, changing the quality of evoked activity, rather than exciting or inhibiting neurons directly. For example, in times of high arousal noradrenaline is released, and this focuses cortical attention on threats. The noradrenaline arousal system can influence large areas of cortex in this way.

Columnar organization

Cortical processing is dominated by local circuits producing columnar activity. The shape of individual cortical neurons shows that most of their connections are made with cells above and below them, which means that cortical activity tends to involve groups of cells arranged in a column. Some cells in each column make inhibitory connections with adjacent columns, that tends to focus cortical activity.

Cortical connectivity

The cerebral cortex is made up of a rich tapestry of interconnections between cortical areas, enabling cooperative analysis of information. On the other hand, the connections between the cortex and non-cortical parts of the nervous system are quite circumscribed. For example, the connection between the thalamus and the cortex is very specific. All sensory input to the cortex (except olfaction) is relayed to the cortex via thalamic nuclei, and each major region of neocortex is connected to a particular nucleus of the thalamus. In return, the cortex regulates thalamic activity through feedback pathways, which are actually larger than the input pathways from thalamus to cortex. To direct the rest of the nervous system, cortical outputs target control systems in the thalamus, hypothalamus, hippocampus, brainstem, and spinal cord, rather than directly activating effectors, like muscles and glands [6].

Maintaining spatial order

A common property of sensory projections is that they retain a spatial arrangement for their entire journey through the nervous system. The spatial (topographical) arrangement results in what is called somatotopy for the touch system, which means that the map of the skin of the body is transmitted to the cortex in proper spatial order. An outstanding example of somatotopy is seen in the representation of rodent whiskers in the somatosensory cortex. Each whisker is connected to a cylinder of cortex called a barrel. In the case of the visual system, the retinal map is maintained (retinotopic organization), and in the case of the auditory system the tonal map is transmitted intact (tonotopic organization). Topographic representation provides the ability to compare activity from adjacent patches of skin, or finding the edges of visual objects.

The cortical column as a basic functional unit?

Professor Vernon Mountcastle suggested that the cerebral cortex is divided into 'mini-columns'–groupings of neurons which work as functional units for cortical processing. Other neuroscientists disagree with this view: although there is good evidence of strong columnar activity in the cortex, it has been argued that the varying size of observable columnar units suggests that there are no consistent units across the cortex. It is possible that some regions have distinct units whereas others form temporary groupings to suit the demands of particular tasks.

---

[6] As Chapter 4 describes, humans and their close primate relatives have many direct connections between cortical neurons and motor neurons, but this is the exception rather than the rule.

Figure 7.4 Barrel fields in the mouse cortex

This picture shows the correspondence of the whiskers of the face of the mouse with specific areas on the opposite side of the cortex. These areas are called barrel fields due to the cylindrical arrangement of cortical cells, with each whisker being represented by one barrel. (Reproduced with permission from Woolsey and Van der Loos, 1970, Figure 15)

Whiskers and barrels

The discovery of special cortical areas (barrels) for individual whiskers in mice by Woolsey and van der Loos in 1969 opened up a whole new field of cortical research. The fact that this striking anatomical feature had remained undiscovered for so long was probably due to the routine use of coronal sections, in which the barrels are hard to see. In sagittal or tangential sections, the rows of barrels, one for each major whisker, are obvious.

The functional layout of the human cerebral cortex

The cortex is relatively uniform at a cellular level, but identification of its different regions has led to an understanding of functional specializations. The earliest attempts to map the human cerebral cortex were based on the patterns of gyri and sulci on the surface of the cerebral hemisphere. The cortical surface was divided into lobes named after the bones that cover them: frontal, parietal, temporal, and occipital. The lateral fissure and the central sulcus assist with the separation of the frontal, temporal, and parietal lobes from each other; the boundary between the occipital lobe and the parietal and temporal lobes is an arbitrary line.

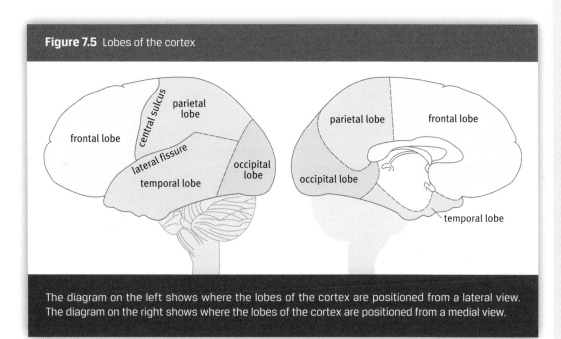

Figure 7.5 Lobes of the cortex

The diagram on the left shows where the lobes of the cortex are positioned from a lateral view. The diagram on the right shows where the lobes of the cortex are positioned from a medial view.

Although the surface shape of the cortex differs between individual humans, some gyri and sulci are consistently present. For example, there are always three parallel gyri on the lateral surface of the frontal lobe; they are called the superior, middle, and inferior frontal gyri. They end at the precentral gyrus, the primary motor cortical area in humans. The lateral surface of the temporal lobe also has three long parallel gyri running from front to back; they are called the superior, middle, and inferior temporal gyri.

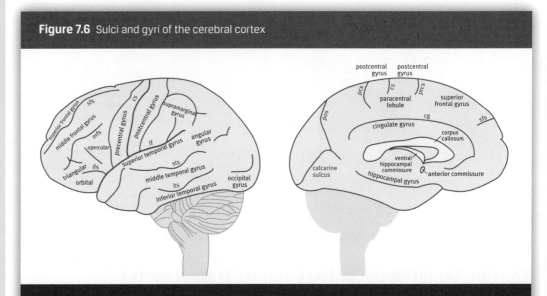

Figure 7.6 Sulci and gyri of the cerebral cortex

The diagram on the left shows the position of the main gyri and sulci of the cerebral cortex from the lateral side. These sulci include the superior frontal sulcus (sfs), middle frontal sulcus (mfs), inferior frontal sulcus (ifs), central sulcus (cs), lateral fissure (lf), superior temporal sulcus (sts), and the inferior temporal sulcus (its).

The diagram on the right shows the position of the main gyri and sulci of the cerebral cortex from the medial side. These sulci include the calcarine sulcus, parieto-occipital sulcus (pos), post-central sulcus (pcs), central sulcus (cs), pre-central sulcus (prcs), superior frontal sulcus (sfs), and the cingulate sulcus (cg).

The pattern of gyri and sulci in human brains

There is widespread variation in the pattern of gyri and sulci in humans, and there have been elaborate attempts to work out the meaning of this variation, including a study of the brains of brilliant people, such as Einstein. The results have been disappointing, as they have not revealed unique patterns in special people. It seems likely that the secret of Einstein's brain may have been in the richness of his synapses, rather than its surface appearance.

Naming the cortical areas

Anatomists of the early 20th century tackled the arduous task of dividing the human cerebral cortex into different areas based on the thickness of cellular layers, cell form, and cell density. Campbell, Brodmann, and von Economo each made excellent maps which subdivided the cortex into distinct regions, but it is Brodmann's map that became the standard reference. The human cortical areas are routinely referred to by the Brodmann numbering system. For example, in Brodmann's system the primary motor cortex is area 4 and the primary visual cortex is area 17.

One limitation of Brodmann's mapping work is that in many cases he examined only the cortex on the surface of a particular gyrus, and did not attempt to map the cortex in the walls of the sulci between different gyri. This criticism does not apply to his mapping of the lateral fissure, the largest of the sulci, because he realized that a number of important areas were buried in the hidden banks of this deep sulcus. Among the areas in the lower bank of the lateral fissure are the auditory area (area 41) and the language area of Wernicke (areas 39 and 40). If the banks of the lateral fissure are pulled apart, a buried island of cortex is revealed in the depths of the fissure. This is called the insula. It is the location of the cortical area which receives taste information.

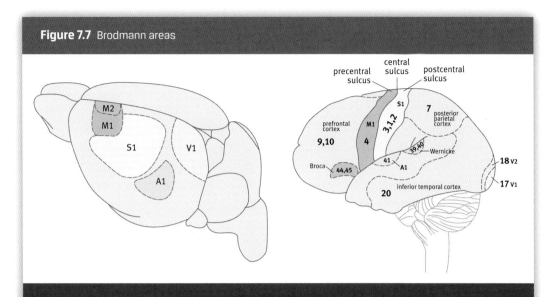

Figure 7.7 Brodmann areas

The diagram on the left is of a rodent brain showing segments of Brodmann's map of the cortical areas. The diagram also shows Brodmann areas in relation to the main functional areas of the cortex; M1 (primary motor cortex), M2 (secondary motor cortex), S1 (primary somatosensory cortex), A1 (primary auditory cortex), and V1 (primary visual cortex).

The diagram on the right is of the human brain showing some of Brodmann's areas. The diagram also shows Brodmann areas in relation to the main functional areas of the cortex; M1 (primary motor cortex), S1 (primary somatosensory cortex), A1 (primary auditory cortex), V1 (primary visual cortex), V2 (secondary visual cortex), W (Wernicke's area), Broca (Broca's area), prefrontal cortex, posterior parietal cortex, and inferior temporal cortex.

Identifying the function of different cortical areas in humans

Identifying the function of individual areas of the cortex in humans is a very difficult task. Most of the information we have about cortical function in humans comes from careful observation of patients with damage to particular cortical regions. In recent decades, these clinical studies have been strengthened by the use of a variety of modern imaging techniques such as fMRI (see Chapter 11). These techniques measure changes in cortical activity or metabolism when people perform tasks, or experience stimuli, but they can only reflect activity that occurs on a relatively large scale.

The use of electrodes to record directly from brain tissue gives much more precise information about cellular activity, but in humans is only used during neurosurgery, or to study electrical activity triggering epileptic seizures. Electrode recording studies in experimental animals, that provide the bulk of our detailed knowledge of cortical function, are of limited usefulness because they are not matched to human cortical abilities. Despite these problems, the location of a range of functions and abilities have been identified, so that we can even identify areas related to face recognition and musical ability.

Each part of the cortical sheet receives a unique mix of inputs from sensory systems and other cortical regions, so its analysis is similarly unique. For example, parts of the parietal cortex receive visual and tactile inputs, enabling an understanding of how objects can be manipulated, and informing decisions about how to use them by influencing motor planning regions. The input mix of each cortical region corresponds to its function, such as Wernicke's region, which receives

---

Misleading maps of auditory and visual cortex

Many textbooks, including this one, are guilty of misrepresenting the position of the auditory and visual cortical areas on pictures of the lateral side of the human brain. The reason for this is that the auditory cortex is almost all hidden in the lateral fissure, and the visual cortex is almost all hidden on the medial side of the occipital lobe. In an attempt to show their position in relation to other cortical areas, it is customary to show both these areas on a lateral view, even though this is actually wrong.

sounds processed by the auditory cortex, and is also strongly connected to the speech motor areas. This combination allows it to analyze sound into the elements of speech, so that spoken words can be decoded and comprehended.

Primary, secondary, and association areas

The function of some cortical regions is obvious–for example, activity in the primary visual cortex can be linked to specific kinds of stimuli, and mapped in great detail using electrophysiological studies of experimental animals. Other regions combine more complex information, and the results of their analysis are difficult to characterize. Damage to the latter regions changes behavior in ways that may take years to become apparent to others. In general, the best understood cortical areas are those that correspond most closely to sensory inputs or motor outputs, called primary regions. Adjacent to primary regions, and strongly interconnected with them, are secondary regions. These secondary regions interact with primary regions to extract meaning or focus on particular aspects of primary activity, such as visual forms or sequences of movement. Most secondary regions are also reasonably well understood, at least in terms of their specializations. The dynamics of how they specialize on those particular qualities are extremely complex.

As we move further away from primary regions, cortical processing is less and less related to identifiable real-world events or responses, and much more oriented to the private, internal communication patterns of the cortex. Very little is known about the form of information communicated between these cortical regions. Current scientific methods can describe the cellular properties and connectivity to a limited extent, but how these arrangements convey meaning is almost impossible to analyze, since it involves the simultaneous firing of millions of cells.

These highly sophisticated areas, termed cortical association areas, make up the bulk of the human neocortex. Some clues to the function of the association areas are offered by the study of some people with cortical damage. These people may lose the ability to combine certain types of information, such as the ability to name objects or read maps, or they may have larger deficits such as a complete loss of color perception. In prefrontal regions, the result of damage may be subtle, such as small deficits in planning, organization, and concern for the future. There may also be inappropriate behavior in social settings. These types of deficits have taught neuroscientists and neuropsychologists a great deal about the processing abilities of the cortex, but we still lack any real understanding of how these abilities are generated.

The cerebral cortex and behavior

Much of the survival behavior in mammals does not primarily involve the cerebral cortex. Most basic behaviors and many learned movements take place with minimal cortical input. This is evident in experiments in which rats have had their cortex removed; the rats are still able to feed, drink, and avoid threats.

However, damage to some cortical areas in humans can produce subtle but profound personality changes and behavioral problems. In the past it was popular to talk about 'silent' areas of cortex but it is now apparent that the function of these areas is simply too sophisticated to be measured in standard clinical situations. The cerebral cortex is extremely expensive to run in terms of metabolic energy, and its cells are in near-constant activity during awake behavior. Any animal with silent cortical regions would simply be wasting energy, endangering its survival.

The main job of the cerebral cortex is to analyze, predict, and respond to environmental events.

Alexander Luria and aphasia

The brilliant Russian researcher Alexander Luria is credited with inventing the field of neuropsychology. During the Second World War he made a detailed study of the abilities of soldiers who had suffered small head injuries. His analysis is the foundation of modern theories on aphasia (the inability to communicate). The introduction of Armalite-style weaponry in recent years means that small wartime head injuries are now extremely rare, because a modern bullet flattens on impact to make much larger wounds. Because of cold-war politics, American scientists largely ignored Luria's extraordinary work until the 1970s.

This allows the cortex to shape our behavior in very sophisticated ways. The cortex is able to learn from previous experience, tying together information from different sensory systems and mental states in order to make predictions about what is likely to happen in the future. The cortex makes use of abstract representations of sensory inputs, which it receives pre-processed from the thalamus. Similarly, the cortical representation of movement is also at a conceptual rather than concrete level. It is more concerned with 'where' and 'how' a movement should happen, delegating the job of contracting specific muscles to the embedded pattern generators and reflex circuits in the brainstem and spinal cord.

Differences between the two cerebral hemispheres

Some people with intractable epilepsy have been treated by surgically disconnecting the cerebral hemispheres from each other to prevent the spread of seizures. This involved the sectioning of the corpus callosum. Such 'split brain' patients have been studied since the 1960s by Michael Gazzaniga and Roger Sperry, among many others. These studies have looked at differences in the processing of the left and right cerebral hemispheres, by controlling what information is received by each side, and asking for responses using the left and right hands, which are independently controlled in these patients. The left brain-right brain dichotomy they postulated has become a favorite topic in popular psychology, but is nonetheless based on well-controlled scientific studies. Broadly speaking, each of the hemispheres has specialized functions. In right-handed people, language and calculation are almost exclusively handled by left hemisphere regions, whereas spatial awareness is strongly correlated with the right hemisphere.

Even when evaluating similar stimuli, the two hemispheres often come up with quite different results. Left hemisphere analyses tend to focus on the component parts of a stimulus, breaking it down into simpler pieces. Right hemisphere analysis is often more concerned with the whole form of the stimulus, disregarding the smaller scales. For example, in a piece of music the left hemisphere may track the melody of individual notes, while the right hemisphere understands the musical key, and the larger themes of the composition as a whole. While listening to speech, the left hemisphere may focus on words and meaning, whereas the right hemisphere listens to the pitch and timing, gaining additional meaning. When the corpus callosum is intact, these alternative analyses are combined, and given different amounts of attention depending on their usefulness in a particular situation.

Language areas in the cerebral cortex

The ability to convey ideas through language (orally or in writing or through signing) is perhaps the most extraordinary feature of the human brain. Two areas in the cerebral cortex have been found to be associated with language. Since speech is the dominant form of language, they are commonly referred to as the speech areas. One of these is in the frontal lobe and is named after the 19th century neurologist Paul Broca, who first associated damage there with speech difficulties. The second speech area, found at the junction of the parietal and temporal lobes, is named after Broca's contemporary, Carl Wernicke. In right-handed people, the two main speech areas are located in the left cerebral hemisphere, but in left handed people the speech areas are either in the right or left hemisphere. The hemisphere which contains the speech areas is called the dominant hemisphere for language. The dominant hemisphere controls the preferred hand for fine manipulations. Since the dominant hemisphere is on the left side in 90% of people, the dominant hand is usually the right hand.

Arcimboldo and the hemispheres

A painting by sixteenth century artist Giuseppe Arcimboldo, showing a face made up of images of vegetables, has been used to demonstrate a major difference between left and right cerebral hemispheres. Dr Mike Gazzaniga showed that when these paintings are shown to a split-brain patient, the right hemisphere reports seeing a face, whereas the left hemisphere reports seeing the objects which compose the face–in this case, vegetables. Reductionist versus holistic analysis is a consistent difference between the left and right hemispheres.

The largest of the two speech areas is Wernicke's area, which is located next to the auditory cortex. The comprehension of spoken or written words takes place in Wernicke's area and the regions that surround it. Wernicke's area is also central to the generation of speech. The speech commands are sent to Broca's area, which converts them into sounds or written words. If Broca's area is damaged, the person may be able to understand speech and think of what they want to say, but the words do not come out. If the person can speak at all, it is through the production of disjointed words with the omission of functional words and inflections. Damage to Wernicke's area produces an even more profound deficit in communication, characterized by an inability to relate ideas to words, and a lack of understanding of spoken or written words.

The name given to the communication problems that arise from damage to the speech areas is aphasia. The type resulting from damage to Broca's area is referred to as expressive aphasia. The type resulting from damage to Wernicke's area and surrounding regions is referred to as receptive aphasia. In receptive aphasia, understanding of speech is severely affected and patients are unable to follow commands or repeat what is heard. In some of these cases, the person may still be able to speak (because Broca's area is intact), but the result is a meaningless string of words described as a 'word salad'. Damage to Wernicke's area and the regions around it may also be associated with alexia (inability to read) and agraphia (inability to write).

Expansion of the cerebral cortex in recent human evolution

The relative independence of the cerebral cortex from the rest of the nervous system has allowed the cortex to alter its size and functional arrangement without disrupting core survival functions. This is clear from the great diversity of cortical sizes among mammals, whose need for cortical function is dependent on their different lifestyles and dietary choices.

While a large cerebral cortex is a prominent feature of primates, the recent rapid expansion of the human cerebral cortex has no parallel. In a period of about four million years, there has been an increase in brain size from about 450 cm^3 in our australopithecine ancestors to about 1,300 cm^3 in modern humans. Almost all of this increase in brain size is related to cortical expansion. It has been suggested that the main driver for this expansion is the increasing social complexity of human populations. This theory argues that the expanded human cortex is related to the increased demand for success in complex social interaction. Over time, this complex social interaction has been a product of high levels of communication and the ability to cooperate with others in very large social groups. Studies of monkey social groups have found that those species with large communities have bigger brains than closely related species that live in small family groups.

The importance of social intelligence is supported by numerous recent studies on social position and self-esteem. Poor self-esteem is associated with chronic stress and serious health consequences. In a large study of British public servants, those in the lowest (fifth) grade suffered premature death from heart attack more than three times as often as those in the first grade. This was not a threshold effect associated with relative poverty; instead, there was a linear increase in death rates from the first public service grade to the second, and so on to the fifth. This indicates that with each drop in status, there was an increase in social stress and a loss of self-esteem. Similar results have been obtained from studies of monkey societies, where chronic social stress has a devastating effect on those at the bottom of the hierarchy.

Norman Geschwind and cerebral dominance

Dr Norman Geschwind was an outstanding neurologist and an inspirational teacher at Harvard Medical School. He developed a simple but brilliant method for identifying anatomical evidence for cerebral dominance in post-mortem human brains. He showed that measurement of the parts of lower bank of the lateral fissure with a ruler revealed the dominant hemisphere in most cases. He was an international leader in research in cerebral dominance, dyslexia, and epilepsy.

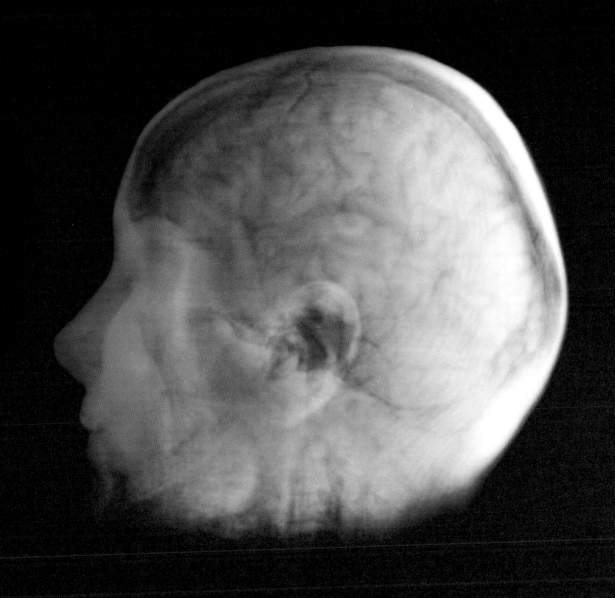

Chapter 8
**Higher level functions
consciousness, learning, memory, and emotions**

8 Higher level functions
–consciousness, learning, memory, and emotions

In this chapter:

Consciousness	110
Memory	111
Sleep	117
Emotions and the amygdala	121

Consciousness

Consciousness represents the highest level of brain functioning, and it has proved extremely difficult to study. To be able to discuss consciousness one must be able to distinguish between conscious and unconscious states, which turns out to be quite complicated. On the one hand, a person can be awake but not fully conscious, and on the other a limited level of consciousness seems to exist in dreams, during rapid-eye-movement (REM) sleep. Disorders such as sleepwalking, night terrors and narcolepsy have demonstrated that the relationship between consciousness and sleep is complex.

There is no single consciousness center in the brain. Many studies have concluded that consciousness arises from the coordinated activity of many parts of the cerebral cortex. A theory of consciousness that has recently become more popular is the 'global workspace' concept put forward by Bernard Baars in 1938. Baars suggested that the separate sensory areas of the cerebral cortex–the visual, touch, hearing, and taste areas–are aware of sensory input, but not at a conscious level. But if the simultaneous activity of many of these centers is broadcast to other areas of the brain, then consciousness could be said to exist. The main areas that respond to the pooled information from the sensory cortical areas are the prefrontal cortex, the cingulate areas, and the parietotemporal association cortex. These areas of cortex are known to be involved in high-level cognitive activity, such as reasoning and planning. The areas that respond to pooled sensory information act as a global workspace, according to Baars' concept. The French team led by Dehaene and Changeux found a massive burst of activity in the global workspace areas (particularly the frontal lobes) 300 milliseconds after the cortex receives an important sensory stimulus. The Dehaene team also found that people with damage to their prefrontal cortex take longer to become conscious of sensory inputs.

The clear implication is that consciousness–the mental experience of nervous system activity–lags behind the processing of sensory events. We become aware of things slightly after they reach our sensory systems, a time lag that may be necessary to assemble sensory information from vision, touch, hearing and other senses into a combined experience. Occasionally we become aware of this lag, such as when withdrawal of the limb from a burning match occurs before we become aware of the sensory events which prompted it.

Two minds in one brain?

Split-brain patient studies provide interesting insights into the pathways involved in consciousness. It has been well established that for a person in whom the two halves of the cerebrum were surgically separated, the left-oriented side of the cortex is unaware of visual sensation on the right side of the visual world. A word or picture flashed to the left visual field will register on the right side of each retina, and hence will be seen by the right hemisphere only. In these cases the left hemisphere (the typical site of the language centers) is incapable of verbally describing the stimulus, but it may use the left hand to point in response to questions about the stimulus. This strongly suggests that some of the information associated with consciousness is usually shared via the corpus callosum, but cannot be transmitted via subcortical pathways. Experiments like these, in which the hemispheres of the cortex make separate responses based on different information, suggest that two independent reasoning processes are occurring in a single individual. It seems that their left and right hemispheres can experience different states of consciousness.

There are many interesting and worthwhile studies of these neural correlates of consciousness, but there are no experiments or theories that satisfactorily deal with consciousness itself. Scientists are unable to make the link between the activity of the brain and the subjective awareness we each experience. As scientists, we can say what processes are necessary or critical to consciousness, but there is as yet no account of how brain activity generates that state of mind.

Memory

The ability to recall previous events and use them to guide future behavior is perhaps the most remarkable ability possessed by the brain. For most animals, crucial memories are related to place, such as where to find shelter, where to find food, and what places are dangerous. Humans have an enormous capacity to remember, ranging from an ability to recall complex music scores to detailed recollections of scenes from childhood.

Three main types of learning and memory may be usefully distinguished. The first is the recollection of specific events and places, called episodic memory by psychologists, and what we usually refer to when we use the word 'memory' in conversation. The second is our accumulated experience, which allows us to understand the world and predict what will happen in a given situation, often termed semantic memory. Thirdly, the nervous system learns and optimizes common patterns of activity, such as the movement sequences of walking or throwing a stone. This type of memory can be described as motor learning. Some people use the term 'muscle memory' to describe this process, but this is misleading because memories of muscle action are functions of the cerebellum, striatum, and cerebral cortex. The memories do not reside in the muscles themselves.

Basic mechanisms of learning and memory

Experience changes the future actions of the nervous system. This is a form of learning. The types of experience that cause these changes, and the molecular processes used, vary considerably between different parts of the nervous system. However, there are some common elements in the types of changes that can be achieved. The main type of change is thought to be in the connectional structure of networks of neurons, which requires alterations to their synapses. Three main mechanisms of synaptic modification have been described.

Firstly, neurons can alter which neurotransmitters are used at the synapse, so that the activity of one cell has a different kind of effect on the other. This commonly happens during development, so that a synapse may start out using glutamate (excitatory) early on, then shift to GABA (inhibitory), and finally to glycine as the connections are worked out. This method seems not to be used in mature nervous systems, whose neurotransmitter usage remains fairly stable after the developmental period.

Secondly, connected neurons can alter the amount and rate of neurotransmitter release on the pre-synaptic side, or the number and type of receptors on the post-synaptic side. Such changes can greatly alter the size and quality of the influence transmitted, and can occur very rapidly as receptors and vesicles are shifted to and from the membrane. This is the most studied kind of learning mechanism, particularly the phenomena known as long-term potentiation, long-term

Eric Kandel and memory mechanisms

Professor Eric Kandel of Columbia University was awarded the Nobel Prize in 2000 for his work on molecular changes during learning. With his colleagues, he discovered important biochemical changes that were associated with short-term memory formation. Despite scepticism from other scientists, this work was carried out on a very simple animal model system. He studied learning associated with reflexes in the marine mollusc *Aplysia*.

depression, and spike-timing-dependent plasticity. Finely balanced biochemical interactions between enzymes, transport proteins, and release of calcium ions allow patterned activity to produce lasting changes in the strength and type of a synaptic connection.

Thirdly, to achieve large-scale alterations in function, cells may disassemble, regrow or physically alter their dendrites, and possibly their axons as well. Recent studies have described how changes in activity lead to the formation and breakdown of dendritic spines, dendritic branches and even entire dendrites as needed. Silencing of synaptic inputs can lead to the neurons losing many of their dendrites, and then growing new ones to receive active inputs. This type of large-scale reorganization, termed plasticity, allows the nervous system to salvage functional arrangements even after quite significant damage or disruption. The degree to which plasticity is possible changes from region to region, and drops off during development and maturation, although even in adult nervous systems, parts of the hippocampus remain highly plastic.

In some parts of the nervous system (such as in the hippocampus and the olfactory bulb), new neurons are also added to replenish populations of cells. The relationship between neurogenesis and memory is currently unclear, although there is evidence that the hippocampus replaces cells that have exhausted their functional capacity for learning.

These mechanisms describe how cells can alter connections, but the meaning of these alterations, and the effect they have on behavior or perception, depends on the large-scale organization of the cells that are changing connections. Limitations in current techniques make it difficult to study large-scale neural processing, and so most of our knowledge about learning and behavior focuses on anatomical structures. The assumption is that the cellular mechanisms described are low-level underpinnings of the large-scale functions of the hippocampus, cortex, and other memory-related systems.

Motor learning

An essential learning function of the nervous system is the automation and optimization of movements. The motor system is controlled by a complex interaction between multiple control centers (see Chapter 5), each of which uses feedback from proprioception and tactile receptors to improve its operation. Genetic and trophic differences during embryonic development mean that muscle and bone structure varies from person to person, and throughout life our muscles and bones change. Fatigue, injury, growth, exercise, arthritis, infections, and the influence of hormones all have profound effects on the strength of muscles and the types of movements we can achieve with them. The nervous system needs to learn the relationships between every muscle and every body part, and to keep that knowledge up to date, so that behavioral plans for movement generated by the cortex can be correctly and efficiently executed.

A great deal of knowledge of the relationships between individual muscles is embodied in the networks and connections of the spinal cord. For example, activating a muscle will automatically inhibit muscles that oppose its actions, by invoking an inhibitory interneuron. These relationships may be simple, as in the inhibition of opponent muscles, or they may involve complex activity across dozens of muscles to compensate for postural shifts and imbalances caused by large movements. The pattern generators and movement modules described in Chapter 5 are shaped and informed by the spinal cord's learning of muscle relationships.

Hebbian learning

In the 1940s, psychologist Donald Hebb reasoned that learning and memory could occur if neurons modified their connections on the basis of activity. His insight, often summarized as "neurons which fire together, wire together" has stimulated a huge body of research on the mechanisms of neuronal plasticity. Hebbian learning is not the same as the phenomena of long-term potentiation and depression which are studied in modern laboratories, but it proved to be an important stepping stone to modern theories of memory and brain function.

At a larger scale of organization, the effect of any planned movement on the position and orientation of body parts needs to be predicted with great accuracy. Our movements are planned by forming a target, generally a point in space that we would like part of our body to occupy, such as a firm-looking piece of ground for the foot to step on. To do this, a huge library of information on movements, spatial configurations, and relationships between body parts needs to be compiled and consulted. This is the job of the cerebellum, which indexes what we learn from previous movements–how to get to organize the movement and how to correct errors in placement, either before they happen or on the basis of feedback.

In addition, many of our behaviors consist of movement sequences that activate body parts or behavioral modules one after the other to achieve a desired action. Throwing, for example, involves gripping, adjusting posture, rotating the trunk and shoulder and straightening the arm in a precise sequence, whose timing needs to be very accurate for an effective throw. These types of compound movements are executed clumsily at first, with the cortical motor system triggering each element one after the other. If the movement is repeated a number of times, the pattern becomes familiar to the striatum and cerebellum. This familiarity is used to enhance the fluidity of the movement, by activating each step in the sequence while the previous step is still in progress. Eventually the movement becomes fluid and efficient, allowing the cortex to concentrate on other details, such as where to throw something, rather than how to throw.

The hippocampus and episodic memory

The area of the brain that initially registers episodic memories is called the hippocampus, a distinctive area of three-layered cortex in the temporal lobe of the human brain. In many small mammals, about 10% of the cerebral cortical volume is occupied by the hippocampus, but in humans the hippocampus is overshadowed by the huge overgrowth of the association areas of the frontal and temporal lobes. Non-mammalian vertebrates, such as birds and reptiles, also have a hippocampus. Their hippocampal structure is not as spectacular as in mammals, but many non-mammalian vertebrates still have prodigious memory capacity. It is claimed that birds that hide seeds and other food can remember up to 30,000 hiding places. During the autumn hoarding season, the hippocampus in these birds actually enlarges by 30%.

The hippocampus is involved in memory systems and seems not to be directly involved in consciousness. But without the hippocampus, an individual will experience severe distortions of consciousness, because without recent memory they are unable to create a context for their surroundings. This is supported by studies of many patients with hippocampal damage, including HM, a patient of William Scoville who has been studied by Brenda Milner. Henry Molaison (HM) underwent a bilateral removal of the medial components of the temporal lobes, including the hippocampus, in an attempt to treat non-responsive, severe epilepsy. After the surgery HM reported his conscious state was abnormal. He felt he was "constantly waking up from a dream" with "everything looking unfamiliar," and was unable to form new memories of his experiences after the surgery. For example, he was greatly surprised and distressed by the ageing of his face in the following decades, to the extent that mirrors were removed from his accommodation.

Structure of the hippocampus

The cortical folding in the mammalian hippocampal region is striking, and it is easy to recognize the major subdivisions in histological sections; these subdivisions are the dentate gyrus, the CA regions, the subiculum, and the entorhinal cortex.

Gaps in experience
During the day, short-term associations formed by the hippocampus keep track of experiences, but some things are best forgotten. For clinical procedures such as colonoscopy, the short-acting drug midazolam can be used to boost GABA receptor activity in the hippocampus, suppressing its normal activity. This prevents the formation of new episodic memories while the drug is active–causing temporary hippocampal suppression. As a result, the patient can co-operate throughout the procedure but retains no memory of it afterward, often reported as a "gap in experience".

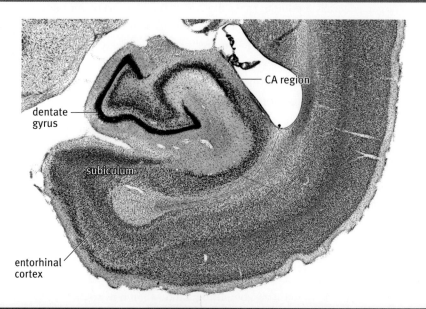

Figure 8.1 Nissl-stained hippocampus

This is a picture of a Nissl-stained hippocampus in a marmoset. This picture shows the distinct folded shape of the hippocampal region. The darkly stained area represents the dentate gyrus of the hippocampus. (Image courtesy of Dr Hironobu Tokuno, Tokyo Metropolitan Institute for Neuroscience)

Rembering places

The hippocampus and the adjacent entorhinal cortex contain many cells that respond to spatial position in the environment. The discovery of 'place' cells in the entorhinal cortex by O'Keefe and his colleagues led to a surge of research on the mechanisms that underlie spatial memory.

The entorhinal cortex helps to link events to places, by registering incoming information on two-dimensional grid maps of the surrounding area–a kind of GPS for the brain. The entorhinal cortex sends this information to the dentate gyrus. The dentate gyrus then projects to the CA3 region, which in turn sends information to the CA1 region. The CA1 region sends its information to the subiculum, the last station in this hippocampal circuit. The subiculum sends its information to septum and hypothalamus via a large fiber bundle called the fornix. This main path (entorhinal cortex-dentate gyrus-CA3-CA1) was named the trisynaptic circuit by Per Anderson. The hippocampus uses this ciruit to code events in relation to specific spatial locations.

The effects of stress on the hippocampus

Stressful situations have a direct impact on the memory functions of the hippocampus. In some instances, the failure to recall a traumatic event may be due to a stress-induced breakdown of hippocampal function. When humans or other animals are stressed, the adrenal gland releases the hormone cortisol into the bloodstream. This release is governed by the paraventricular nucleus of the hypothalamus. The paraventricular nucleus receives inputs from various structures, such as the amygdala. The amygdala stimulates cortisol release while the hippocampus inhibits it, so blood cortisol level is balanced by the influence of these two regions.

If stress persists for too long, the ability of the hippocampus to control the release of stress hormones is compromised, and its ability to perform routine memory functions is reduced. Chronically stressed rats are unable to learn and remember how to perform behavioral tasks that

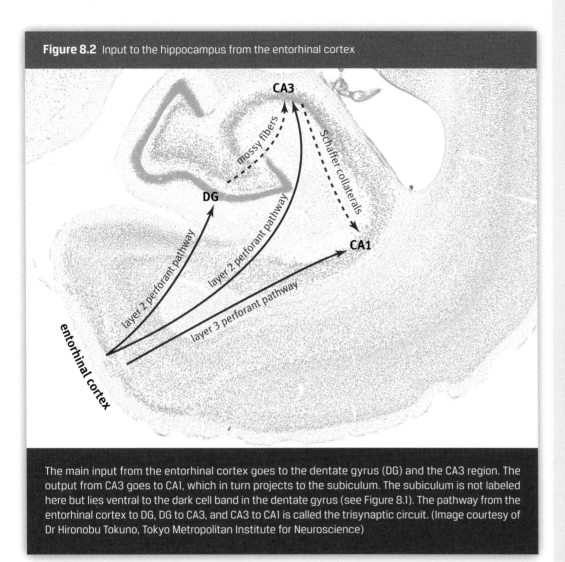

Figure 8.2 Input to the hippocampus from the entorhinal cortex

The main input from the entorhinal cortex goes to the dentate gyrus (DG) and the CA3 region. The output from CA3 goes to CA1, which in turn projects to the subiculum. The subiculum is not labeled here but lies ventral to the dark cell band in the dentate gyrus (see Figure 8.1). The pathway from the entorhinal cortex to DG, DG to CA3, and CA3 to CA1 is called the trisynaptic circuit. (Image courtesy of Dr Hironobu Tokuno, Tokyo Metropolitan Institute for Neuroscience)

The Morris water maze

This test measures the ability of a rat or mouse to learn and remember. The animal is placed into a small pool of water that contains an escape platform which the animal cannot see, because it is hidden just below the water surface. The animal swims around the pool until it finds the platform. On subsequent trials the time to find the platform (latency) decreases, showing that the animal can remember where to find it. The maze is commonly used as a test of hippocampal function.

depend on the hippocampus. Their ability to remember is classically tested with the Morris water maze. Stressed rats are unable to remember the location of the safe platform in a water maze, whereas they are able to do this in a non-stressed state. Chronic stress and high cortisol levels can also disrupt sleep cycles, further interfering with memory. Plasticity in the hippocampus is impaired by chronic high cortisol levels. However, plasticity in the amygdala may be enhanced by sustained stress. The result is a suppression of episodic memory in the hippocampus, but an opportunity for the amygdala to learn from sustained stress.

Severe but temporary stress can result in loss of dendrite branches in the hippocampus. These changes are reversible in the short-term, but if stress is prolonged the dendrite loss becomes irreversible and hippocampal neurons will eventually die, causing permanent impairment of memory. In survivors of chronic severe trauma, such as repeated child abuse, hippocampal volume is reduced in magnetic resonance scans. These people exhibit significant deficits in memory without any loss in IQ or other cognitive functions.

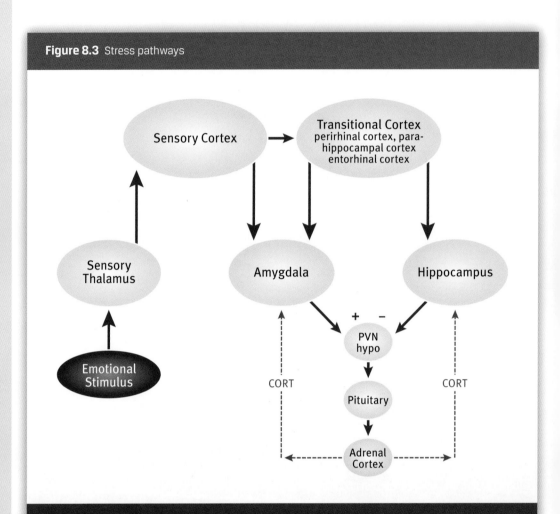

Figure 8.3 Stress pathways

This diagram shows the impact of stress on different parts of the brain. Stress inputs into the hypothalamus activate the endocrine system to produce the stress hormone cortisol (CORT). (Adapted from Le Doux, 1998, p 241)

Stress and the brain

The response of the brain to stress involves the hypothalamus, the pituitary, and the adrenal cortex. The hypothalamus triggers the release of adrenocorticotropic hormone (ACTH) from the anterior pituitary, which in turn causes cortisol to be released from the adrenal cortex into the bloodstream. Chronic severe stress is associated with illness and premature death.

The role of the cerebral cortex in episodic and semantic memory

The hippocampus is crucial for memory formation, but it is believed that in the longer term, episodic memories and semantic memories are stored over a much wider area of the cortex. The ability of the cortex to remember underlies our knowledge and understanding of the world, allowing patterns to be recognized and details to be filled in when they are missing. In this way, most of the connections of the cortex are created on the basis of experience. Early on in development, the basic structure of the cerebral cortex is guided by chemical gradients and genetic programs (see Chapter 10). Primary sensory regions are organized by incoming activity, and this creates the basic layout of these cortical areas.

The large and elaborate human cortex can make fine distinctions and accumulate detailed knowledge of the way the world works. This enables us to make accurate and useful predictions about the results of planned behaviors. The recollection of experience can be classified as either semantic or episodic. Semantic memory relates to a generalized impression of a number of experiences, such as the movement of falling objects, the texture of water, and the meaning of a pattern of sounds. Semantic memory enables understanding rather a recollection of specific experiences. Episodic memory is linked to experiences that occurred in particular times and places. The distinction between the two forms of memory is not always clear-cut. For instance we may know something to be true (semantic), but in the process we may recall a learning experience that taught us that fact (episodic). Furthermore, any piece of knowledge will depend on many episodic experiences, but in the long term the semantic memory remains when the individual episodes are forgotten.

In storing our working knowledge of the world, the cortex needs to adapt and change on the basis of new experience, as well as retaining information that is current and important to survival. The process of updating is thought to be an important activity during sleep, which in effect disconnects the cortex from sensory and motor systems, in order to update and restructure it.

Sleep

Sleep is essential for all vertebrates. However, some other mammals sleep very differently to humans, with periods of sleep between three and twenty hours a day, and varying levels of alertness, such as sleeping half the cortex at a time. We often think of sleep as a way to rebuild energy, but the body and the brain continue to use energy while we are asleep. Very deep sleep may be a way to conserve resources at times when the central nervous system and intelligent behavior are not required. For example, some animals hibernate for much of the winter.

In humans, sleep runs in 90-minute cycles. In each cycle, the proportion of slow-wave activity in the electroencephalograph (see Chapter 11) gradually becomes greater, eventually reaching deep sleep. This cycle is followed by short periods in which brain activity resembles an awake state, called rapid eye movement (REM) sleep. Subjects woken from various stages of sleep report that dreams occur during REM periods. On the other hand, the mental states of deep sleep are vague and not dream-like. In addition to marked changes in the electrical activity of the brain, the activity of various neurotransmitter systems changes from deep sleep to REM sleep.

REM sleep is characterized by high levels of cortical acetylcholine (ACh), reflecting a high level of activity and arousal. High ACh levels also prime genetic mechanisms that underlie synaptic plasticity in the cortex. During slow-wave sleep ACh levels are much lower, permitting the activation of hippocampal information, and the consolidation of that experience into the structure of the cortex, locking in new knowledge. These processes can be blocked by raising ACh levels during slow-wave periods, and this impairs learning. Informally, we might think of dreaming as testing how cortical knowledge deals with recent events, and deep sleep as an update of the cortex on the basis of that testing.

What regulates sleep?

The regulation of sleep is the function of a handful of cell groups in the hindbrain and hypothalamus. These areas have widespread, diffuse projections to the cerebral cortex that are responsible for keeping us awake and alert. These 'awake' centers are the locus coeruleus, the serotonin raphe system, the dorsal tegmental nuclei, and the tuberomammillary nucleus of the hypothalamus. A special feature of this group of nuclei is that each relies on a different transmitter substance–noradrenaline in the locus coeruleus, serotonin in the raphe, acetylcholine in the dorsal tegmental nuclei, and histamine in the tuberomammillary nucleus. Sleep occurs when these centers are shut down.

The identification of noradrenaline, dopamine, and serotonin nuclei

Until the 1960s, only the acetylcholine transmitter could be identified in brain sections. A breakthrough came with the development in Sweden by Falck and Hillarp of a method which identified catecholamines in the brain by causing them to fluoresce. The Swedes were able for the first time to identify the location of noradrenaline, dopamine, and serotonin cell groups in the brainstem.

Changing names: the reticular activating system

The centers to keep the brain awake were originally described by physiologists as the reticular activating system of the brainstem. A variety of experiments showed that ascending pathways from the brainstem caused the cerebrum to be alert. Physiologists hypothesized that the brainstem nuclei that gave rise to the ascending pathways reside in the so-called reticular core of the brainstem. The was no anatomical evidence to support this. This caused confusion over the identity of the brainstem centers that promoted wakefulness, and only recently have the specific nuclei been identified. Since the nuclei that promote wakefulness (the locus coeruleus, the raphe nuclei, and the major cholinergic nuclei) are not part of the true reticular formation of the brainstem, the term 'reticular activating system' should be replaced by 'ascending arousal system'.

Staying awake

Ascending arousal systems keep the cerebral cortex active in part by stimulating the thalamus, and axonal projections from the cortex to the thalamus can perform a similar role. If the cortex has a need to remain very active, it can drive thalamic nuclei to keep itself awake, although this becomes increasingly difficult when ascending arousal systems are being suppressed by the VLPO.

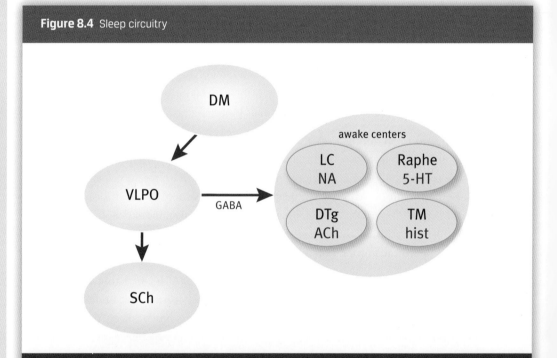

Figure 8.4 Sleep circuitry

This diagram shows the brain pathways involved in sleep control. Inputs from the visual pathway (light or dark) influence the preoptic area. The preoptic area then sends inhibitory signals to the awake centers. For example, night influences the preoptic area to inhibit awake centers, producing sleep. The diagram also shows how inputs from the dorsomedial nucleus can aid in regulating awake and sleep. DM=dorsomedial hypothalamic nucleus; SCh=suprachiasmatic nucleus; LC=locus coeruleus; NA=noradrenaline; 5-HT=serotonin; DTg=dorsal tegmental nucleus; ACh=acetylcholine; TM=tuberomammillary nucleus; hist=histamine.

The neurologist von Economo suspected that some part of the brain had the ability to switch off the 'awake' systems. Working in Vienna during the viral encephalitis epidemic of 1905, he was able to examine the brains of people who had died from three different types of sleep disorder. The patients suffered from intense sleepiness, insomnia, or narcolepsy.

In the group with intense sleepiness, von Economo found lesions near the junction of hypothalamus and midbrain, and he suspected that there was damage to a brainstem center or ascending pathways, which kept the brain alert. In the group with insomnia, he found lesions in the anterior hypothalamus and preoptic area, and suggested that there had been damage to a center that makes the brain go to sleep. In the group who had suffered from episodes of falling asleep uncontrollably (narcolepsy), he found lesions in the posterior half of the hypothalamus, and speculated there was damage to a center that served to stabilize the relationship between sleep and wakefulness.

In the 1990s, the Saper group at Harvard began an ambitious project to unravel these pathways. By tracing the input connections of some of the awake centers, they discovered a master sleep switch rostral to the hypothalamus in the preoptic area. The sleep switch is a small cell group called the ventrolateral preoptic nucleus (VLPO). The VLPO makes inhibitory connections with all the 'wake' centers (locus coeruleus, raphe, the cholinergic tegmental nuclei, and the histamine nucleus). When the VLPO is activated, it switches off the wake centers, and the brain goes to sleep. Damage to the VLPO produces pathological insomnia, a permanent inability to sleep. The activity of VLPO is principally controlled by input from the visual pathway. Specialized cells in the visual system connect to the suprachiasmatic nucleus, a 24-hour time-keeper. This nucleus tells the VLPO when it is dark, and a good time to go to sleep, and when it is time to wake up.

The final link in the sleep circuitry is the loop that stabilizes the system. If the brain had only reciprocally interconnected centers for sleep and alertness, there would be a tendency for behavior to flip unpredictably between sleeping and waking, as happens in narcolepsy. The dorsomedial nucleus of the hypothalamus has a stabilizing role, continually monitoring activity and behavior, and recognizing when danger or other factors make sleep inappropriate, even when VLPO has calculated it is time to sleep.

Histamine and the brain

A single nucleus in the brain uses histamine as a transmitter substance. It is the tuberomammillary nucleus of the hypothalamus. The nucleus projects to many areas of the forebrain and plays a role in arousal. Older-style antihistamine medications cause sleepiness by blocking the effects of this nucleus.

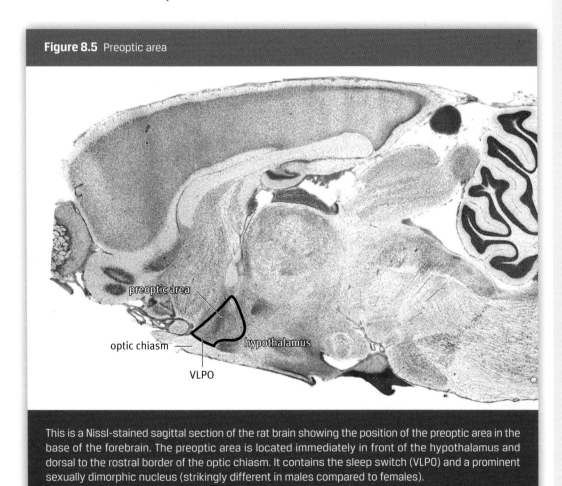

Figure 8.5 Preoptic area

This is a Nissl-stained sagittal section of the rat brain showing the position of the preoptic area in the base of the forebrain. The preoptic area is located immediately in front of the hypothalamus and dorsal to the rostral border of the optic chiasm. It contains the sleep switch (VLPO) and a prominent sexually dimorphic nucleus (strikingly different in males compared to females).

Brain clocks–the systems that control behavioral states

Most mammals, including ourselves, have a 24-hour cycle of wakefulness and sleep. This cycle of brain activity is controlled by a small number of centers in the brain, which operate as biological clocks. The technical term for a 24-hour cycle of sleep and wakefulness is circadian rhythm. The 24-hour clock is crucially important because it correlates with the external cycle of night and day, but the brain also has other clocks that operate on shorter periods–from one to three hours.

The main 24-hour clock resides in a pair of very small cell groups next to the midline of the anterior hypothalamus. Because they sit just above the crossing (chiasm) of the optic nerves, they are called the suprachiasmatic nuclei. Although the cells of the suprachiasmatic nuclei have their own intrinsic clock mechanism, they also receive information about light and dark from the nearby optic nerve, and this allows them to synchronize or entrain the operation of the clock. If a person is kept constantly in complete darkness, the suprachiasmatic clock slips into to a 25-hour cycle, but it switches back to 24 hours once the person returns to daylight.

The rate of secretion of many hormones is controlled by brain clocks. The rate of secretion varies over the 24-hour cycle, with peaks and troughs at different times of the day. The 24-hour clock also regulates some brain functions that have a longer periodicity, such as the timing of ovulation in the menstrual cycle of women. The timing of ovulation has been intensively studied in female rats, which have a four- or five-day cycle of ovulation. Ovulation is controlled by the secretion of gonadotropin releasing hormone (GnRH), which causes the anterior pituitary to release gonadotropic hormones (FSH and LH), which in turn act on the ovary to secrete estrogen. The hypothalamic cell group that secretes GnRH (called AVPe) receives signals from the 24-hour clock in the suprachiasmatic nucleus and somehow converts this into a four- or five- day signal for the timing of GnRH release.

Rhythm generators in the hindbrain

The 24-hour clock for the brain is in the suprachiasmatic nucleus of the hypothalamus, but the brain also has clocks with shorter cycle times. The fastest clocks are in the caudal hindbrain, where a number of centers generate rhythms that underlie important functional systems. These include the respiratory rhythm pacemaker (the pre-Bötzinger nucleus), the vocalization rhythm generator, and the cerebellar timing system (the inferior olive).

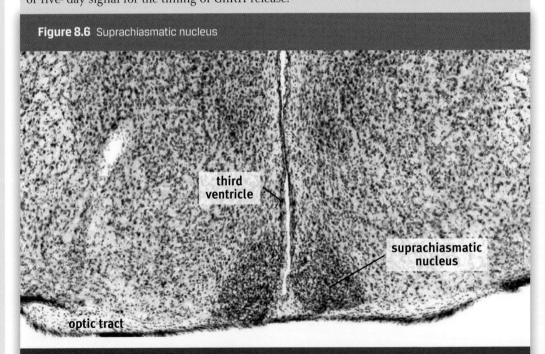

Figure 8.6 Suprachiasmatic nucleus

This is a Nissl stained coronal section of a mouse brain showing the darkly stained suprachiasmatic nucleus on both sides of the third ventricle. The suprachiasmatic nucleus is a prominent cell group in the most rostral part of the hypothalamus, sitting just above the optic chiasm. It receives a direct input from the retina and acts as a 24 hour clock for the brain.

Emotions and the amygdala

Given our relatively sketchy understanding of how behavior is created and regulated, the concept of emotion is quite difficult to address. Emotions may be thought of as global behavioral patterns, since the kinds of behavior we expect from someone who is angry would be quite different from someone who is sad in the same situation. The effects of different emotional states are also obvious in facial expressions, posture and even in the style of movements we make. We say that a happy person has a spring in her step, whereas an angry person may stalk off in a rage. We are exquisitely sensitive to these emotional cues because they allow us to interpret and predict the behavior of others, but many of them are relatively subtle nuances and therefore difficult to study. Emotions that involve high levels of physiological arousal, such as stress or fear, are experimentally accessible, and so we know more about them. In stressful or highly emotional behavior, the organizing centers appear to be in the amygdala.

The amygdala is a cluster of nuclei located in the temporal lobe of the cerebrum. The amygdala can be divided into two main groups: the cortical/medial group and the lateral/basal group. The two groups have very different anatomical and functional relationships. The cortical/medial group is connected to the olfactory system, whereas the lateral/basal group is involved in complex circuits relating to emotional responses, particularly fear.

The two diagrams below are based on those in an excellent review of the anatomy of the amygdala by Joseph LeDoux. They summarize the input and output relationships of the main parts of the amygdala. The lateral/basal amygdala is responsive to sensations that relate to danger, and it can react by stimulating the autonomic and endocrine centers, which prepare the body for action in the face of danger. Most of these responses are activated through connections of the amygdala with the hypothalamus.

What scares us

The alarm functions of the amygdala serve to focus attention and behavior on threats that should not be ignored. There is evidence that the amygdala triggers fear behaviors when a snake-like object is seen, even in captive monkeys that have never seen a snake before.

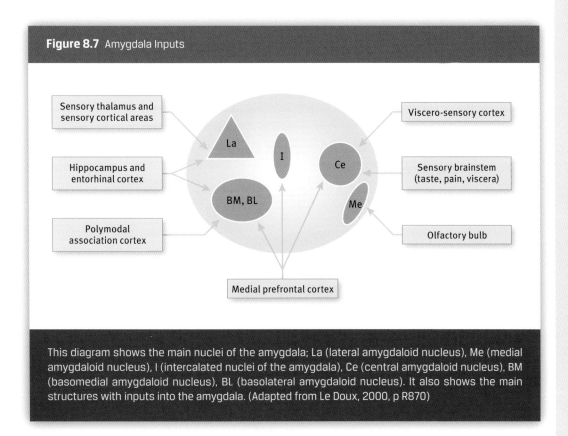

Figure 8.7 Amygdala Inputs

This diagram shows the main nuclei of the amygdala; La (lateral amygdaloid nucleus), Me (medial amygdaloid nucleus), I (intercalated nuclei of the amygdala), Ce (central amygdaloid nucleus), BM (basomedial amygdaloid nucleus), BL (basolateral amygdaloid nucleus). It also shows the main structures with inputs into the amygdala. (Adapted from Le Doux, 2000, p R870)

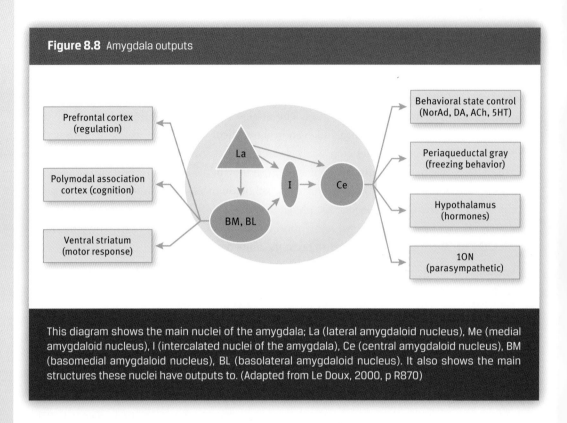

Figure 8.8 Amygdala outputs

This diagram shows the main nuclei of the amygdala; La (lateral amygdaloid nucleus), Me (medial amygdaloid nucleus), I (intercalated nuclei of the amygdala), Ce (central amygdaloid nucleus), BM (basomedial amygdaloid nucleus), BL (basolateral amygdaloid nucleus). It also shows the main structures these nuclei have outputs to. (Adapted from Le Doux, 2000, p R870)

Neuroscientists Heinrich Kluver and Paul Bucy, at the University of Chicago, found that removing the amygdala of rhesus monkeys had a significant effect on aggressive tendencies and responses to fearful situations. The surgery produced numerous behavioral abnormalities, now called the Kluver-Bucy syndrome. After surgery, monkeys appeared to have good visual perception but poor visual recognition. The most dramatic emotional changes present were a reduction in aggression and fear. This was evident as a decrease in both the experience and expression of fear and aggression, with the monkeys having a placid demeanor, even in the presence of a natural enemy such as a snake. Stress has a negative effect on the ability of the hippocampus to register new memories, but it does not damage the amygdala, and may actually enhance amygdaloid function. It is therefore possible that an individual may have lost conscious memory of a traumatic event but at the same time they may retain very powerful, unconscious emotional memories through amygdala-mediated fear conditioning. These unconscious fears may become resistant to extinction and consequently become unconscious sources of intense anxiety.

The amygdala and social hierarchy

It has often been suggested that the amygdala might play an important role in hierarchical behavior (pecking order) in vertebrates. A recent study on people who had suffered damage to the amygdala found that they lost their sense of personal space. This is consistent with the amygdala having a role in human social behavior.

Understanding the relationship between the amygdala and aggressive behavior in animals is challenging, because the nature of aggressive behavior is often misinterpreted. Konrad Lorenz, winner of the Nobel Prize in 1973, has pointed out that a lion attacking a springbok is not being aggressive; it is just securing a meal. He argues that the most significant aggressive behavior is that which occurs between members of the same species, usually in an attempt to exert dominance over others in a hierarchy. It may be that the amygdala drives aggressive acts that establish the social position of an individual in a group.

Changing names: the 'limbic system'

In many textbooks, a chapter dealing with the amygdala and the hippocampus would be titled 'The Limbic System.' For many reasons, we recommend you avoid this confusing term.

The term limbic system has its origin in 'le grand lobe limbique' of Broca, who drew attention to the curved belt of telencephalic structures that border the central parts of the forebrain. This definition of the 'limbic lobe' includes the hippocampus, the amygdala, the cingulate gyrus, and the fornix. In the contemporary literature, the limbic lobe is often said to include the various parts of the hippocampal formation, the septum, the olfactory tubercle, the bed nucleus of the stria terminalis, the amygdala, and the cingulate gyrus. Different authors add or subtract structures from the definition in a random manner. As Brodal critically observed, "It is difficult to see that the lumping together of these different regions under one anatomical heading, 'the limbic lobe' serves any purpose."

As Brodal also points out, the use of the term 'limbic system' is even more dangerous, because the supposed components do not share a common function. Instead, each has a distinct set of functions, ranging from olfaction to emotional responses and memory registration. In an attempt to deal with the diversity of structures thrown in to the 'limbic system' Mesulam and colleagues (1977, p 407) have suggested the definition of 'limbic' should include all centers that are either directly or indirectly connected with the hypothalamus. This is similar to the attempts to make the term 'limbic system' equivalent to the now discredited term 'visceral brain.' However, there are many other areas of the brain that are involved with emotion and the autonomic nervous system, that are not part of the limbic system as typically defined. Conversely, the hippocampus is involved in memory and cognition but not in emotion.

The answer to these increasingly confusing attempts to defend an unfortunate name is to abandon the term 'limbic' altogether. Instead, one should speak specifically about the hippocampus, amygdala, septum, or whatever part of the brain is under discussion, without connection to the baggage created by the term 'limbic.'

Drives and rewards

The function of the nervous system is to enable survival, and the ability of the vertebrate brain to do this under different conditions has shaped its evolution. Part of the success of complex nervous systems is their ability to continually impel the individual to behave in a purposeful manner. To do this, the nervous system drives behavior in a way that is linked to internal systems of motivation.

Some of these drives are very direct: when we are thirsty we look for water; when we are hungry we look for food. In these cases, the drive is immediately satisfied by getting the water or the food. But running in parallel with these direct behavioral drives are systems of internal motivation and reward. Brains have chemical systems that are activated when the animal successfully achieves a goal. Once the goal is achieved, the animal feels a sense of relief and pleasure, and will be more likely to pursue goals in the future. A potent region for reinforcing goal-oriented behavior is the ventral tegmental area (VTA) in the midbrain and adjacent regions. The neurons of the VTA send dopaminergic projections to certain parts of the forebrain. This fiber system, which terminates

---

Pecking order

Pecking order is the name commonly given to the dominance hierarchy in animals. The phenomenon was first described in chickens almost a hundred years ago, but it is an important feature of social organization in many animals. A higher position in the hierarchy gives an animal preferential access to food and mates. It has been suggested that the hierarchy minimizes conflicts between members of a group. Social hierarchies in human societies have a powerful effect on self esteem, and this raises stress levels in individuals at the bottom of the hierarchy.

mainly in the nucleus accumbens, is called the medial forebrain bundle (mfb). The VTA-mfb projection has been called the 'reward system' of the brain. The activation of the nucleus accumbens is thought to signal to the cortex that something worthwhile has happened, and subsequent neural activity in the ventral pallidum and orbitofrontal cortex is associated with feelings of pleasure. A number of addictive drugs activate these pathways artificially, producing the feelings of pleasure and satisfaction in the absence of goal oriented behavior. This leads to behavior that is targeted at repeating the drug effect, which is the basis of addiction.

We tend to think of own behavior as existing in two different compartments; one that is driven by what we consider to be free will, and another that is driven in a direct or reflex way by particular kinds of stimuli, over which we have little control. However, every behavior is a blended combination of volitional, cortical-type behavior planning, integrated with the direct, automatic circuitry that produces a predictable response pattern under given conditions. Pinning down the difference between free will and behavior driven by needs or stimuli is difficult, since we often feel 'ownership' of drive-motivated behavior like obtaining food.

Understanding these internal reward systems can be very valuable. People who consciously set goals on a regular basis do so because they realize that they feel happy when they achieve these goals. The goals can be large, like running a marathon, or small, like repairing a piece of furniture. But in both cases, the actions trigger the internal reward systems to bring pleasure and satisfaction. This leads to a sense of happiness that reinforces the behavior, and makes it more likely to be repeated in the future.

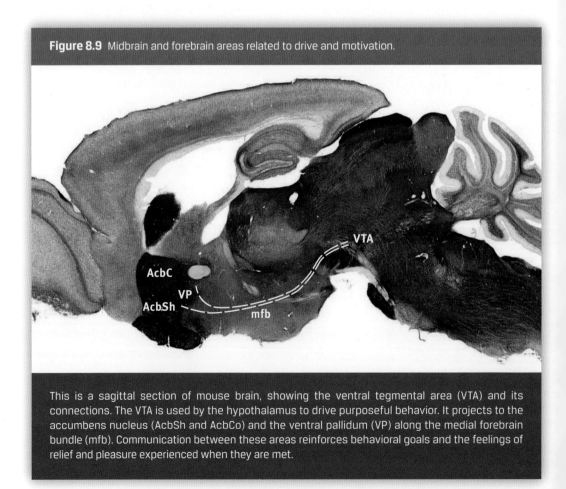

Figure 8.9 Midbrain and forebrain areas related to drive and motivation.

This is a sagittal section of mouse brain, showing the ventral tegmental area (VTA) and its connections. The VTA is used by the hypothalamus to drive purposeful behavior. It projects to the accumbens nucleus (AcbSh and AcbCo) and the ventral pallidum (VP) along the medial forebrain bundle (mfb). Communication between these areas reinforces behavioral goals and the feelings of relief and pleasure experienced when they are met.

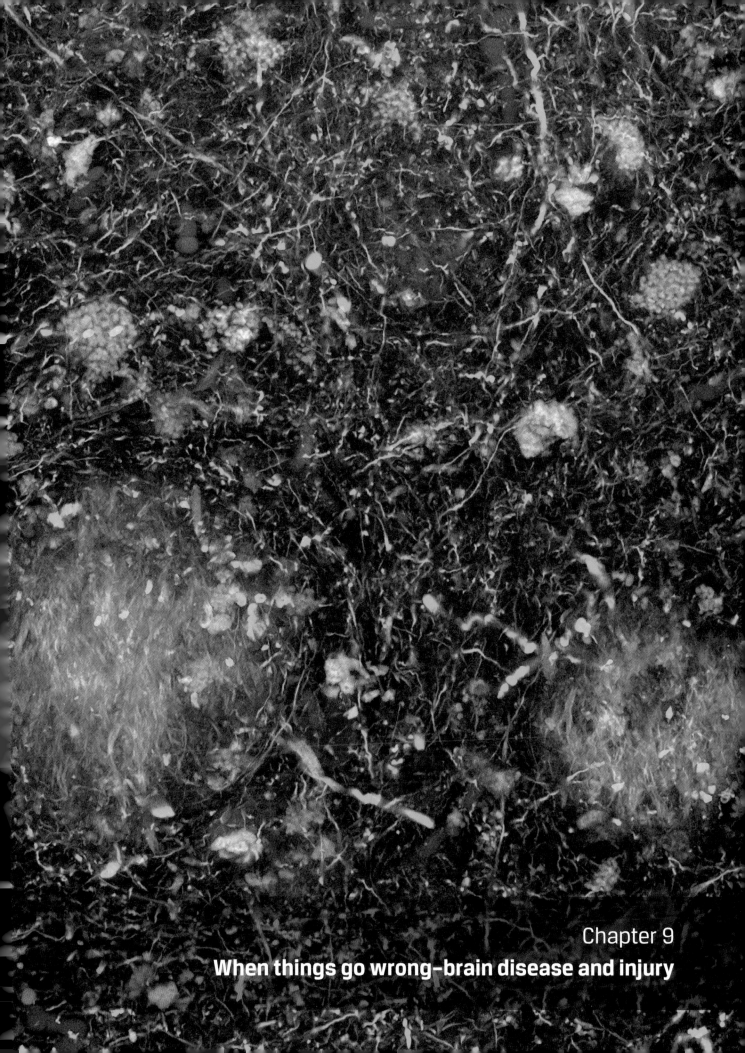

Chapter 9
When things go wrong–brain disease and injury

9 When things go wrong
–brain disease and injury

In this chapter:

Infections of the brain and spinal cord 126

Multiple sclerosis 128

Parkinson's disease 129

Stroke (cerebrovascular accident) 130

Alzheimer's disease and dementia 130

Epilepsy 132

Brain trauma and brain death 132

Mental illness 133

The tragic history of the treatment of severe mental illness 136

When neurons in the brain or spinal cord are damaged by injury or disease, the result is usually very serious. There are two reasons for this. First the neurons and their axons form the command and control system for the body, and damage may interfere with these crucial messaging systems. If some cells in a muscle are damaged, there may be some loss of function, but the effect will be contained to the region of damage. On the other hand, damage to an area as small as a walnut in one cerebral hemisphere may paralyze the whole opposite side of the body.

The second reason why CNS damage is often more serious than in other systems is that once neurons are lost, they generally cannot be replaced. In other parts of the body, such as the liver or kidney, damaged cells can be replaced by new cells and function is maintained. It was previously thought that there was absolutely no replacement of damaged neurons in the brain, but recently some interesting exceptions have been found. However, it can still be assumed that almost all the neurons that die from injury or disease in the brain or spinal cord will never be regenerated. Current research with stem cells aims to solve this problem by inserting cells that are capable of forming replacement neurons.

This chapter will summarize the features of some important causes of damage to the brain and spinal cord, as well as some disorders of mood and behavior whose symptoms are less easily pinpointed. Where possible, we will identify the anatomical correlates of the neurological illness.

Infections of the brain and spinal cord

Poliomyelitis and rabies

Because the brain and spinal cord are so well protected from invasion, infections are uncommon. In rare cases virus particles enter through the blood stream, as in encephalitis. In the case of poliomyelitis (polio), the virus enters via the nerve-muscle synapse, and travels along motor neuron axons to reach the spinal cord or hindbrain. Polio virus is a cause of summer diarrhea in most cases, but when the virus invades the central nervous system it can cause severe paralysis. During the epidemics of the 1950s, hundreds of thousands of people around the world were paralyzed by polio infections every year.

Once the polio virus reaches the motor neuron cell body, it causes fatal damage, which results in paralysis. Unfortunately, the damage is permanent because the motor neurons never regenerate. If the motor neurons affected are in the spinal cord below the mid-cervical region, the result is paralysis of one or more of the limbs. If the motor neurons affected are in the hindbrain and/or the upper cervical region, the result is paralysis of the swallowing and breathing muscles. In the latter case, the person will die unless their breathing is supported by a mechanical respirator. During the polio epidemics of the 1950s, these respirators were huge metal containers that surrounded the body, and were called iron lungs. The discovery and widespread use of protective vaccines has gradually eliminated polio from almost all countries in the world, and in 2010 only four countries still report polio cases on a regular basis.

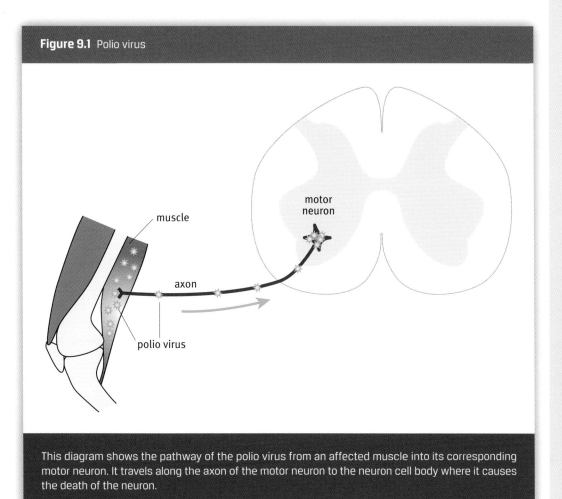

Figure 9.1 Polio virus

This diagram shows the pathway of the polio virus from an affected muscle into its corresponding motor neuron. It travels along the axon of the motor neuron to the neuron cell body where it causes the death of the neuron.

Rabies is another virus that uses peripheral nerves as a way to enter the central nervous system. Rabies virus is present in the saliva of infected bats and dogs. When a person is bitten or scratched by an infected animal, the virus travels along axons and enters the spinal cord. The special feature of the rabies virus is that once it reaches cells in the spinal cord, it can cross synapses to reach all the neurons that connect with the infected neuron. In this way, the virus can slowly spread to virtually all the cells in the spinal cord and brain. Eventually, sometimes years after the original bite, the cells begin to die and death inevitably follows. This terrible process can be interrupted if a special vaccination is administered within a few days of the initial bite.

The ability of rabies virus to cross synapses has been used in the development of nerve tracing techniques that use pseudorabies virus, a non-dangerous relative of the true rabies virus.

Mad cow disease and CJD

Mad cow disease and Creutzfeldt-Jakob Disease (CJD) are both examples of a strange neurological disease that can be transmitted from one individual to another but is not caused by a bacterium or a virus. The transmissible agent is a prion: a misconfigured version of a normal protein from the body, which can cause normal proteins to change to its abnormal form. In CJD, rogue prion molecules spread abnormality in the new host, eventually causing widespread brain damage and death. CJD appears spontaneously in about one in a million people. The characteristics of prion disease were unravelled during the study of a kuru epidemic in the highlands of Papua New

---

Tracing with pseudorabies virus

Pseudorabies virus is taken up by terminal boutons and transported in a retrograde direction along the axon to the cell body of a neuron. The virus has the capacity to enter the boutons, which synapse with the infected neuron. In this way, the virus can be followed through a chain of neurons, and can therefore be used to trace multisynaptic neuronal pathways.

Guinea. Kuru, a variant of CJD, causes severe cerebellar and cerebral damage in humans, and resulting in death approximately one year after symptoms commence. Kuru was intensively investigated by Carlton Gadjusek and his colleagues, who showed that the kuru agent, later identified as a prion, could be transmitted to chimpanzees in the laboratory. They showed that the epidemic was caused by ritual cannibalism of deceased relatives in the affected community, and that eating the prion was enough to infect the nervous system. The epidemic ended with cessation of ritual cannibalism in the 1970s, and Gadjusek was awarded the Nobel Prize in 1976.

A variant of CJD called scrapie was thought to affect only sheep, but a similar disease appeared in cattle in Britain in the early 1980s. It is most likely that scrapie was transmitted to cattle by the addition of sheep abattoir by-products to stock feed supplements that were given to cows. The result was an epidemic of bovine spongiform encephalopathy (BSE) in cattle. BSE is the bovine equivalent of scrapie and was named mad cow disease. It was first thought that BSE could not be transmitted to humans eating meat from afflicted cattle, but this belief was tragically wrong, and many people were infected in the UK. In humans, the BSE prion produced a disease described as variant CJD (vCJD). Beef products were withdrawn from sale in the UK and thousands of infected animals were slaughtered and burned. Despite the eradication of the disease from farmed cattle, new human cases of vCJD continue to appear. The reason for this is that prion diseases can have very long incubation periods, perhaps 20 years or more.

Multiple sclerosis

Multiple sclerosis (MS) is a progressive and degenerative disease of the nervous system resulting from CNS inflammation and demyelination of axons. MS is one of the most common causes of neurological disability that predominately affects young adults. MS is three times more common in young women than in young men. Inflammation and damage to myelin sheaths is the result of an autoimmune reaction. This occurs when the immune system of the body becomes deregulated and begins to attack and kill other cells in its own body (auto meaning self). The result of this inflammatory reaction is the proliferation of astrocytes to form scars around the outside of axons. These scars are called plaques. It is the appearance of these plaques that is a characteristic sign of multiple sclerosis.

Nerve impulses normally travel rapidly along the an axon by jumping between gaps in the myelin called nodes. In MS, demyelination and the formation of plaques interrupts the passage of nerve impulses along the nerve axon. The slowing of nerve impulses is the basis of the symptoms experienced by patients with multiple sclerosis. Patients commonly experience abnormalities in motor control or loss of vision. Classic features include muscle weakness, lack of coordination, intention tremor and even paralysis. Multiple sclerosis also causes sensory abnormalities. The optic nerve is commonly affected in MS, resulting in blurred vision or blindness. Immense fatigue is also a common and debilitating symptom. Additional symptoms such as heat sensitivity and electric shock-like sensations are also experienced by some patients. The symptoms of multiple sclerosis usually follow a characteristic pattern, in which episodes of damage are separated by months without new symptoms. The symptoms that arise from each episode are caused by inflammation, and they often gradually resolve as the inflammation in the nervous system settles. Over time, the episodes of damage become more frequent and there is an overall gradual decline in the health of the person. In rare cases, symptoms may progress rapidly, without resolving, until the patient becomes completely incapacitated.

Prions

Prions are proteins normally found in the brain. Their function is not known but they may have a role in memory formation. In CJD and related diseases, the prion proteins become misfolded and accumulations of the misfolded form will damage and kill neurons. The misfolded proteins have the capacity to replicate themselves in a new host by taking over the machinery of the cell.

Parkinson's disease

Parkinson's disease (PD) occurs in older adults. The main features are tremor in the hands, a shuffling walk, and muscular rigidity, making it difficult to initiate a movement. There is no paralysis, but the rigidity in many muscles makes it hard to carry out any voluntary movement, except very slowly. Because of this the patient may appear to be emotionless and apathetic.

We do not know the real cause of PD in the majority of cases, but we do know that the underlying damage is most often in the substantia nigra. The substantia nigra sends dopaminergic axons to the forebrain, mainly to the striatum (caudate and putamen). In PD, most of the dopamine cells of the substantia nigra die, disrupting its normal role in posture and movement.

For many years, the main treatment of PD has been the administration of a dopamine precursor to increase the production of dopamine in the remaining substantia nigra neurons. The chemical precursor is called L-dihydroxyphenylalanine (L-DOPA). This treatment can reduce the severity of symptoms in many cases, but after some years it becomes less effective because too many neurons have died to make effective use of the L-DOPA. Intensive research has focused on the possible use of stem cells to replace substantia nigra neurons, but results have so far been patchy.

In drug-resistant patients, surgical approaches were developed to block the increased tone and tremor in muscles by surgically interrupting the striatal output through the thalamus (stereotaxic thalamotomy). In stereotaxic thalamotomy, a surgeon inserts a special probe into the brain to reach the relevant thalamic nucleus. Because the brain contains no pain receptors, this procedure can be carried out while the patient is awake, which enables the surgeon to see whether the probe is in the right place. This treatment, while reasonably effective in the past, has been superseded by another surgical approach of recent years, deep brain stimulation. In deep brain stimulation, electrodes are implanted to continuously stimulate certain brain areas and reduce the severity of parkinsonian symptoms. Unlike thalamotomy, this technique can be reversed by turning off the stimulator. The main target has been the subthalamic nucleus in the region at the junction of the thalamus and hypothalamus, but more recently the adjacent zona incerta has emerged as an alternative target.

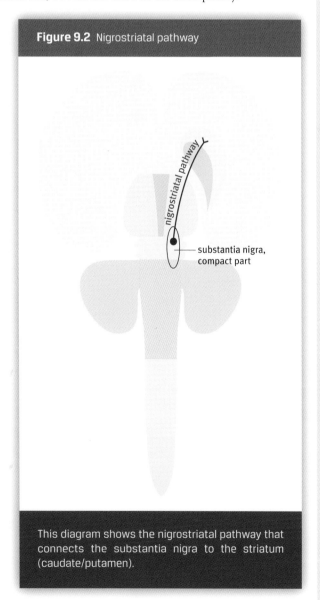

Figure 9.2 Nigrostriatal pathway

This diagram shows the nigrostriatal pathway that connects the substantia nigra to the striatum (caudate/putamen).

Why is the substantia nigra black?

In fresh brain tissue the substantia nigra is colored black, because it contains significant quantities of the pigment neuromelanin, which is closely related to the skin pigment melanin. The chemical transmitter dopamine is manufactured in the substantia nigra, and small amounts of neuromelanin are formed as a by-product of dopamine production. In Parkinson's disease, dopamine synthesis is impaired and the substantia nigra loses its black color.

Stroke (cerebrovascular accident)

The term 'stroke' refers to a situation in which a person is suddenly struck down by something that happens in their brain. The most common cause is bursting or blockage of an artery in the brain, which is described as a cerebrovascular accident, or CVA. Depending on which artery is affected, the result might be paralysis, loss of touch sensation, loss of the ability to speak, or all of these. Because each cerebral hemisphere controls muscles on the opposite side of the body, the paralysis is on the opposite side to the arterial hemorrhage or blockage. In cases where there is extensive damage to the left cerebral hemisphere in a right-handed person, there is often an inability to speak (aphasia) as well as paralysis and sensory loss. The aphasia may be due to damage to Broca's area. In some cases of stroke, the damage may be very localized (such as paralysis of the leg muscles alone), but in cases of massive arterial hemorrhage the damage may involve most of one cerebral hemisphere. In these very severe cases, the brain swelling may press the hindbrain respiratory centers against the skull, resulting in loss of breathing, and death.

One area of the cerebral hemisphere that is particularly sensitive to damage is the internal capsule. The internal capsule is the great fiber bundle that connects the cerebral cortex with the rest of the brain and spinal cord. Even relatively limited damage to this bundle can be catastrophic, in the same way as cutting a major telephone cable can affect a whole office block.

Muscle tone and stroke

Major strokes cause paralysis on the opposite side to the hemorrhage or thrombosis, but they also cause changes in muscle tone. Many muscles become partly contracted and stiff (known as spasticity) and the tendon reflexes are exaggerated on the affected side (hyper-reflexia).

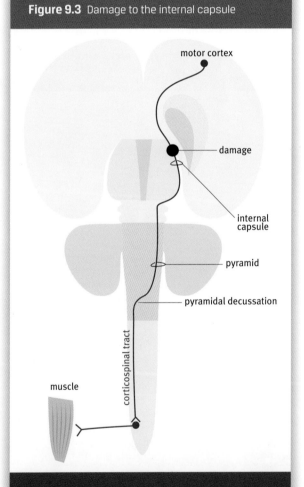

Figure 9.3 Damage to the internal capsule

This diagram shows the corticospinal tract traveling from the motor cortex to a motor neuron supplying a muscle on the opposite side of the body. Because the motor fibers converge in the internal capsule injury to this area may cause substantial paralysis. These fibers are yet to cross the midline, hence this paralysis occur on the opposite side of the body to the area of the brain that is damaged.

Alzheimer's disease and dementia

Alzheimer's disease (AD) causes progressive and relentless loss of nerve cells in the forebrain. The earliest symptoms may be minor episodes of memory loss, but as the disease progresses, the incapacity becomes overwhelming and the person is unable to care for themselves. This state of severe loss of cerebral capacity is called dementia. AD is not the only cause of dementia–in some cases the cause is gradual blockage of the cerebral arteries or Parkinson's disease. While AD is much more common in older individuals, dementia is not a natural result of ageing, and those who escape AD may function at a high level into their ninth decade.

At present there is no way of halting the progress of AD, and therapy is focused on the management of symptoms. Despite intensive research on the underlying causation of AD, the results have been disappointing. We know that there is a genetic predisposition in a small number of cases, and there are indications that the disease may be tied in with loss of insulin receptors in the brain, or damage to connectivity between neurons. One feature of AD is a reduction in the level of acetylcholine in the cerebral cortex, and a number of drugs aimed at sustaining acetylcholine levels have been developed. While these drugs do have a measurable effect, their overall impact is minor and the cost is high.

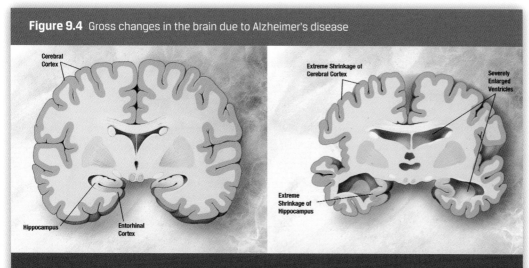

Figure 9.4 Gross changes in the brain due to Alzheimer's disease

This diagram shows the differences between a normal brain (left) and that of a person suffering from Alzheimer's disease (right). The cerebral cortex in the Alzheimer's brain is shrunken because of loss of neurons. The cavities of the lateral ventricles enlarge to take up the space.

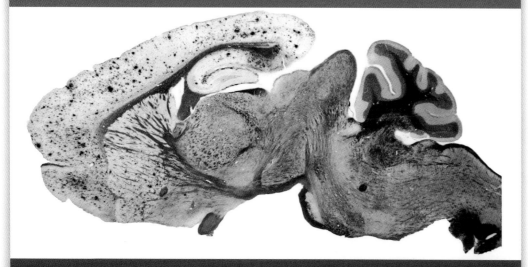

Figure 9.5 Cortical plaques in a mouse model of Alzheimer's disease

This is a section of the brain of a mouse that is suffering from a condition similar to Alzhiemer's disease. The section is stained with the Campbell-Switzer method which shows the presence of plaques (black spots) in the cerebral cortex. The diencephalon and brainstem are not affected at this stage. (Image courtesy of Dr Bob Switzer, Neuroscience Associates, Knoxville TN).

Tau proteins and Alzheimer's disease

The tau proteins play an important role in normal neuronal function. In Alzheimer's and some other neurodegenerative diseases, the tau proteins become attached to a large number of phosphate molecules (hyperphosphorylation) and form abnormal protein aggregations, called neurofibrillary tangles.

Epilepsy

The cerebral cortex usually maintains a close balance between excitation and inhibition of neurons, because of the high degree of interconnection between cortical regions. When this balance fails, the most common result is hyperexcitation, which causes epilepsy. In epilepsy, over-reactive regions stimulate each other and cause 'storms' of electrical activity to spread over the cortex. Since the regulatory mechanisms for balancing excitation and inhibition are unknown, the causes of epilepsy are difficult to discern in many cases. In some cases, an obvious focal cause can be detected and treated with surgery. In the majority of cases people with epilepsy are treated with a combination of drugs that alter the actions of excitatory or inhibitory neurotransmitters. These effects are felt by the entire cortex, so that function in non-epileptic regions is impaired to some extent. These side effects are frustrating for patients and clinicians alike, and the search for optimum benefits with minimal impairment can involve dozens of drug combinations.

Temporal lobe epilepsy (TLE) is a particular form of epilepsy that is characterized by abnormal behavior. TLE can be triggered by scarring from areas of damage to the medial side of the temporal lobe. These injuries can occur during a difficult birth, when pressure on the brain pushes the tip of the temporal lobe against a ridge of dura mater. Later in life, these areas of scar tissue trigger electrical disturbances, which spread through the temporal lobe. Because aggressive behavior is a sometimes a symptom during TLE attacks, many of the sufferers were in the past sent to prison or to psychiatric hospitals. During the 1960s, it was recognized that this syndrome was simply a form of epilepsy and that it could be successfully treated with anti-epileptic drugs.

TLE mainly affects structures located on the medial side of the temporal lobe, including the hippocampus, uncus, and the amygdala. The study of patients with TLE may give us clues as to the functions of these complex areas within the temporal lobe. For example, patients frequently experience abnormal sensations of smell when having a temporal lobe seizure, which we can reasonably attribute to the uncus. The uncus is the site of the primary olfactory cortex. Disturbances of memory function have been attributed to abnormal function of the hippocampus, and aggressive symptoms have been attributed to alteration of function of the amygdala. During the first stage of a TLE seizure, the person may experience an aura made up of olfactory hallucinations, feelings of déjà vu or jamais vu, feelings of fear or terror, and feelings of depersonalization. As the seizure area spreads out to affect a wider area of the temporal lobe, the person often loses consciousness or appears to be in a dream-like state. They may perform abnormal repetitive movements such as lip smacking, chewing, tooth-grinding, or aggressive actions.

Brain trauma and brain death

The brain is protected by the thick bones of the skull, but severe blows to the head can cause irreversible damage to the brain, even if the skull is not fractured. Heavyweight boxers and football players who have experienced repeated concussions during their sporting career have been found to have suffered many brain hemorrhages, with widespread loss of cortical neurons. At first the brain can compensate, but in time the individuals experience memory losses, headache, and confusion. During the recent conflicts in Iraq and Afghanistan, many soldiers have suffered long-term brain damage from shock waves from road-side bombs. These cases were at first dismissed as 'shell shock', but scans have revealed major levels of brain damage. In these cases, there are no broken bones, but the delicate brain tissue suffers damage from the shock wave associated with the explosion. One insidious form of trauma-related damage is diffuse axonal

---

Amyloid

The word "amyloid" means starchy; it refers to the gluey consistency of the insoluble deposits found in brains of individuals who have dies from Alzheimer's disease. It is unclear as to whether amyloid deposits are actively toxic to neurons, or whether they are formed as a by-product of other pathological processes.

---

injury, the gradual degeneration of fibers in the brain due to relatively mild twisting and stretching during head trauma. This type of degeneration can take months to compromise function, so that symptoms appear long after the acute injuries have healed.

Intracranial pressure can rise dangerously if there is a hemorrhage in the brain (such as occurs in stroke) or when infections cause brain tissue to swell. In motor vehicle crashes, the brain trauma may be so great that there is considerable swelling of the cerebrum. Brain tumors in the skull expand slowly, which at first can be compensated by a reduction in the amount of cerebrospinal fluid (CSF). However, once the tumor reaches a certain size, it occupies too much space and begins to raise intracranial pressure, which usually manifests as debilitating headaches, as well as nausea caused by pressure stimulation of brainstem vomiting centers.

Situations in which intracranial pressure becomes raised can be life-threatening. The skull and the dura mater form a tight, inelastic enclosure around the brain, and elevated pressure in this space is transmitted downward to the cerebellum and hindbrain. In the most severe cases, the hindbrain is forced down through the foramen magnum (the exit hole from the skull where the hindbrain joins the spinal cord). Because the hindbrain is cone-shaped, it eventually becomes squeezed against the bone. This cuts off the blood supply to the respiratory centers in the hindbrain, and the result is respiratory failure and death. It is ironic that damage to such a tiny area of the brainstem can cause death, whereas surgical removal of a whole cerebral hemisphere is not life threatening.

Mental illness

Many neurological diseases are reasonably well understood in terms of the specific regions of the brain that are affected. Many of these conditions are incurable at present, although some progress is being made toward understanding their neural bases. However, in the case of mental illness, the situation is much less promising. Mental illness affects about one third of people at some point in their life. At present, therapy for mental illness is largely symptom-focused. While the treatment of mental illnesses has improved a great deal in the past 50 years, there has not been much progress in the understanding of the underlying pathological changes that cause these conditions.

The two major types of mental illness are psychoses and neuroses. Psychoses are mental states in which there is a loss of contact with reality. Sufferers experience hallucinations, delusions, disordered thought patterns, and may exhibit bizarre behavior. Neuroses cause depression, distress, and anxiety but the sufferers do not experience a detachment from reality and their behavior is not outside socially acceptable norms. Both psychoses and severe neuroses can be equally disabling.

Schizophrenia

Schizophrenia is a psychotic illness characterized by abnormalities in the perception or expression of reality. It is a severe and frequently chronic illness that often manifests itself during late adolescence. Schizophrenia affects approximately 1% of the population and is the single largest cause of admissions to mental hospitals in Western countries. It also accounts for the largest proportion of permanent residents in such institutions, because it often results in severe levels of impairment and personality disorganization. Individuals with schizophrenia exhibit a wide range of symptoms including disturbances in thinking, hallucinations, and delusions. They may

---

Schizophrenia is not multiple personalities

The word schizophrenia means 'split mind' and there is a popular misconception that it causes a split personality syndrome. In fact, the 'splitting' is between reality and the internal thinking processes (delusions and hallucinations) of the person with schizophrenia. Multiple personality disorder is a rare condition that has nothing to do with schizophrenia.

---

experience flattened affect (decreased emotional responsiveness), apathy, and withdrawal from society. Schizophrenia has an impact on every aspect of the life of the sufferer, and the condition places a severe burden on their family and friends. There has been a great deal of research into the particular brain regions and neurotransmitters involved in the disease, but we still do not have a clear picture of the underlying pathology. Drug treatment is moderately effective in some cases. Some of these drugs work by influencing the level of neurotransmitters such as serotonin and dopamine, but not all sufferers respond to these drugs.

Bipolar affective disorder

Bipolar affective disorder describes a category of mood disorders defined by the presence of one or more episodes of abnormally elevated mood. These elevated states are clinically referred to as mania, hence the original name 'manic depression.' It is common for individuals who experience manic episodes to also experience depressive episodes or symptoms. As with schizophrenia, bipolar affective disorder greatly affects the lives of patients and their families. A number of pharmacological and psychotherapeutic techniques have been applied to the treatment of bipolar affective disorder, but the mainstays of treatment are mood stabilizers such as lithium chloride (which affects intracellular calcium handling) or sodium valproate (which affects neurotransmission). As with schizophrenia, the neurological changes that occur in bipolar affective disorder are still only vaguely understood. Consequently, while the medication may offer symptom relief for a period of time, such treatment does not offer full recovery and most patients will experience further psychotic episodes throughout their lives.

Neuroses

Individuals with neuroses experience extremes of thinking, and the resulting conditions make it difficult to live a normal life. The symptoms are therefore exaggerations of normal mental states and processes, including anxiety, depression, phobias and obsessions. Those who suffer from neuroses do not experience delusions or detachment from reality, but their symptoms can be quite disabling.

Neurotic depression

Neurotic depression is a state of low mood and lack of interest in activity. Some signs and symptoms of neurotic depression are changes in appetite or sleep, loss of energy, decreased libido, irritability, feelings of worthlessness, and thoughts of suicide. Most cases of depression are classified as neuroses, but in severe cases the condition must be considered a psychotic illness, because it is associated with delusional thoughts. Significant depression affects a third of the population at some point in their lifetime, and it generally begins in early adulthood. Hospital data reveal a higher incidence among women, but this may represent a greater willingness of females to seek help. There is no single cause of depression, but rather a host of contributing factors, including family history, trauma, stress, and psychological problems. Neurotransmitter balance is thought to be the baseline of depression, an imbalance that may be induced by environmental events. Despite numerous brain imaging studies, the exact neurotransmitters involved in depression are still not known. Current pharmacological treatment is largely aimed at increasing the function of the neurotransmitter serotonin in the brain, either by decreasing uptake or increasing production. A number of different drugs are currently prescribed, with varying

---

Meditation and depression

The use of meditation for the treatment of mild depression has become increasingly popular ever since the early 1990s. A practice called 'mindfulness meditation' has performed well in clinical trials. Part of the reason for the popularity of this approach is related to increasing scepticism over the use of drugs for mild depression.

efficacies and side-effect profiles depending upon the patient. The real value of pharmacological treatment in mild-to-moderate cases of depression has been questioned, and there is a great deal of current interest in a number of cognitive-behavioral therapies. Electroconvulsive therapy and transcranial magnetic stimulation have also been used to alleviate depressive episodes with varying degrees of success.

Obsessive-compulsive disorder (OCD)

OCD is a neurotic disorder characterized by obsessions or compulsions or both. Obsessions are recurring, intrusive thoughts that invade a person's consciousness and produce anxiety. Common obsessions include worries about contamination from germs, thoughts about committing violent acts, and constant doubt. Compulsions are repetitive behaviors or thoughts triggered by anxiety. Compulsions accompany obsessions in about 80% of cases and are usually simple acts such as washing, checking, or counting. OCD affects between 2% and 3% of the general population and occurs equally in males and females. The precise areas of brain dysfunction have not yet been found. Selective serotonin reuptake inhibitors (which are used in the treatment of depression) have been found to markedly reduce the symptoms of OCD in about 60% of patients. The success of this drug has led to the suspicion that OCD arises primarily from defects in the brain's neurochemical functioning, defects that can arise on their own or be induced by social or other circumstances. Surgical lesions in specific parts of the cingulate cortex have proved beneficial. Deep brain stimulation and vagus nerve stimulation have also resulted in benefits to the patient.

Autism

Autism is a neurodevelopmental disorder that begins in childhood. It affects males four times as often as females. It is characterized by a lack of emotional responsiveness. It is not a distinct clinical syndrome but rather a spectrum of conditions (including Asperger's disorder) in which social, communication, and cognitive skills are disordered. Autistic children frequently do not make eye contact or smile when interacting with others. Autism is also characterized by repetitive movements such as obsessively lining up toys or objects, or making hand-flapping movements. Children with an autistic disorder commonly over-respond or under-respond to sensory stimuli. For example, a child may ignore someone talking to them, but may respond to another sound in such a way that they become distressed and cover their ears. A good appreciation of the impact of autism on a family can be gained from the sensitive novel by Mark Haddon about a boy with autism. The book is called *The Curious Incident of the Dog in the Night-time*.

Causes of autism

There is a genetic component in the causation of autism, but the major predisposing factors are not known. In 1998 an English surgeon named Wakefield published a claim that certain vaccinations were related to autism, and this resulted in a dangerous fall in vaccination levels in the United Kingdom. His work has been discredited, and in 2010 he was struck off the medical register in the United Kingdom for medical and scientific misconduct.

Research into the underlying brain abnormalities in autism has not so far provided real leads to the cause. Symptoms such as emotional difficulties suggest that the amygdala may have a role in this condition. An important finding in autistic patients is reduced activation of the facial recognition area of the temporal lobe. This has important implications for emotional interpretation and communication in autistic children. It is interesting to note that despite difficulties in some areas, individuals with autism frequently have normal intelligence and may even have superior cognitive skills. Many autistic children have exceptional skills in memory, mathematics, and art. The presence of these extraordinary abilities is described as savant syndrome.

The tragic history of the treatment of severe mental illness

The human brain is immensely complex, and it is not surprising that we have struggled to understand mental illness. Fear and the stigma of mental illness resulted in centuries of brutal and arbitrary treatment of sufferers. The First and Second World Wars triggered acute mental disorders in many veterans, and doctors were desperate to find definitive treatments that might replace a lifetime of misery. One of the results of this desperation has been the willingness to seize on novel ideas that seemed to offer a solution to the problem of serious mental illness. These ideas were not scientifically validated, but somehow people were convinced that they were worth trying. Just as we no longer share medieval beliefs about witchcraft and devils, we need to critically examine popular ideas about mental illness and test them scientifically.

Over the past century, a number of attractive paradigms for the treatment of mental illness have been put forward and adopted in Western countries for want of something better. Some, like Freudian psychoanalysis, have been merely expensive and misleading, but others, like frontal lobotomy and Cotton's surgery for septic foci, have been destructive and even homicidal. In the case of psychosurgery, the USA came close to an Orwellian scenario when Richard Nixon's administration planned to use brain operations to deal with 'difficult' prisoners. The history of bizarre treatments for mental illness makes compelling reading; the books by Valenstein (1973, 1986) and Scull (2006, 2007) are particularly valuable.

Freud and psychiatry

Sigmund Freud was a brilliant doctor who is perhaps best remembered for the recognition of child sexual abuse and the discovery of local anesthesia. Freud's theories on personality and psychotherapy became immensely popular, but in the end have not been sufficiently validated to justify their continuation. The problem is that they were based on a very small sample of patients and that there is no systematic evidence of efficacy. In the 1950s and 1960s, many medical schools taught psychiatry from a Freudian point of view, but the number of adherents has dropped rapidly in the past forty years. Freudian therapy was not dangerous in the same way as psychosurgery, but many commentators believe it was an expensive and largely unproductive diversion in the field of mental illness.

Psychosurgery

Historically, one intervention considered for a wide range of psychiatric disturbances has been surgery on the brain. Increasing knowledge of the functions of different brain regions led to a series of well-intentioned but hopelessly crude surgical interventions, which irreversibly damaged the brains of thousands of patients and left many dead or impaired for life.

Frontal lobotomy

Having learned of the results of frontal lobe removal in chimpanzees, Egas Moniz, a Portuguese neurologist, began performing operations in the 1930s to try to alleviate agitated depression. Their initial procedure aimed to damage the corticothalamic connections of the frontal lobe by alcohol injection in severely anxious patients. The procedures were successful to the extent that many symptoms disappeared after surgery, but many of the patients were left with an enduring frontal lobe syndrome, in which there were changes in personality, and a loss of initiative and spontaneity. Moniz's published results, which focused heavily on the positive outcomes, led to a rapid adoption of similar techniques in Western countries.

Freud and cocaine

Freud became addicted to cocaine, but he recognized the potential value of cocaine as a local anesthetic. He was responsible for encouraging the first use of cocaine in eye surgery by Koller in 1884. This was a major breakthrough in eye surgery, but Koller took all the credit for the idea, and Freud's role was ignored. Cocaine is still used as a local anesthetic for cataract removal and other eye operation.

Frontal lobe surgery in effect disconnected most of prefrontal regions from the rest of the brain. While the actual surgical procedure has many variants, it is generally referred to as prefrontal lobotomy. Despite the fact that there has never been a systematic attempt to collect evidence of effectiveness, it became more and more popular in the 1940s and 1950s. The result was the widespread use of this barbaric attack on the brain, which deprived most victims of their most personal cortical functions.

The worst excesses of frontal lobe surgery were inflicted on hundreds of thousands of soldiers who survived World War II, but suffered serious mental illness. Two American psychiatrists, Freeman and Watts, enthusiastically took up the method in the post-war period, using lobotomy for a wide range of conditions. Their procedure, in which much wider parts of the frontal white matter were completely severed, was performed using flat knives entering each side near the temple and then being swung up and down to cut the frontal white matter.

In his telling analysis of the history of lobotomy, Valenstein (1973) estimates that over half a million lobotomies were performed on US World War II veterans. A single team of neurosurgeons of this era (led by Freeman and Watts) is estimated to have conducted over 20,000 lobotomies. Their methods, which were more rapid and destructive than the original procedure, were met with enthusiasm by a portion of the psychiatric community, who saw it as an answer to the ballooning problems of accommodation in the mental health system. Freeman went so far as to invent a quick and brutal version of the procedure that could be performed in a family physician's office under local anesthetic, by driving a sharp instrument through the top of each eye socket, then sweeping it back and forth to sever the tracts. The simplicity and convenience of this 'transorbital leucotomy' meant it could be performed quickly by doctors with minimal surgical training. By 1955 over 40,000 people in the USA alone had undergone this procedure. It is embarrassing to report that Moniz was awarded the Nobel Prize in 1949, a fact that underlines the fallibility of even our most respected scientific bodies.

The widespread introduction of moderately effective pharmaceutical approaches in the 1950s and 1960s made these surgical techniques increasingly unpopular. When the extent of the damage induced and the degree to which it had been misapplied became apparent, psychosurgery of this kind was roundly condemned, virtually ceasing in the early 1960s. Psychosurgery does still exist today, but only for a small range of conditions, such as obsessive-compulsive disorder, in which very localized surgery has been shown to be effective.

Amygdalectomy

A further abuse of psychosurgery, which persisted beyond the demise of frontal lobotomy, was the use of amygdalectomy in humans. Amygdalectomy is the destruction of some or all of the amygdala. Neuroscientists had reported that monkeys in whom the amygdala was destroyed became much more passive. There was speculation that amygdalectomy might be used to deal with aggression in humans. In the Nixon era, the Justice Department in the USA was attracted to the idea that psychosurgery might solve the problem of prisoners convicted of violent crimes. The officials contemplated the possibility of using a new technology to deal with large numbers of African-American prisoners, who were perceived as a threat to white society. The Justice Department funded the leading neurosurgeons, Ervin and Sweet, to set up a large pilot project, with a view to introducing amygdalectomy across the United States. Fortunately, the exposure of this ghastly plan in Valenstein's book Brain Control (1973) led to calls for a halt to psychosurgery in the USA and other Western countries.

Phineas Gage

In 1848, an American railroad foreman named Phineas Gage had a large iron bar driven through his head by an explosion. He survived, but lost almost all of his left frontal lobe. He developed significant changes in personality and behavior. It is commonly said that this case somehow led doctors to develop psychosurgical approaches for mental illness, but there is no real evidence for such a connection.

Henry Cotton and the septic focus theory of psychiatric disorders

Henry Cotton was a well-known US doctor of the 1920s who believed that infections in the teeth, tonsils, stomach, bowel, and uterus were the main causes of mental illness. There was no evidence for this, but like Moniz and Freud, Cotton was a convincing promoter of his theory. He directed a New Jersey psychiatric hospital and removed organs from thousands of patients. Incredibly, although the mortality was about 30%, no one in the hospital tried to stop him. In fact, Cotton won mainstream approval for his theory and his medical colleagues were happy to ignore the many deaths that resulted from his surgery. However, some doctors were brave enough to demand a formal investigation. The review was carried out by Dr Phyllis Greenacre. Her findings exposed the dreadful results of Cotton's butchery of his patients, but her report in 1925 was shamefully ignored by many and was discredited by Cotton's colleagues. Cotton continued his operations for a decade after the report, and killed hundreds more patients.

Delgado and the bull

The ease with which attractive ideas can command scientific and public attention is wonderfully shown by the use of the film of José Delgado and the control of a bull with a radio device connected to brain electrodes. During the 1950s and 1960s, Delgado was professor of neuropsychiatry at Yale University. He expounded extraordinary right-wing ideas about the use of psychosurgery as a tool of political control. Valenstein records that he proposed "a program of psychosurgery and political control of our society. The purpose is physical control of the mind. Everyone who deviates from the given norm can be physically mutilated."

In a much-viewed short film, Delgado was seen to stop a charging bull by using a radio device to stimulate electrodes embedded in the brain of the bull. The location of the electrodes was never verified, and the massive voltage delivered may in itself have been paralyzing, but the image of the angry bull being halted by brain manipulation had a power that went well beyond science. Still frames from this film appear in many textbooks of psychology, thus perpetuating the myth that Delgado created. The idea that violent behavior could be controlled by electrodes in the brain or psychosurgery was one factor that led to the investigation of amygdalectomy for prisoners in America.

Electroconvulsive therapy

The most controversial non-drug therapy still used in psychiatry is electroconvulsive therapy (ECT). ECT is the process of passing a series of electric pulses through electrodes on the scalp to produce widespread depolarization in the neurons of the brain in an anesthetized individual. The result is a burst of seizure-like activity across the cortex, producing unconsciousness and whole-body muscle spasms that result from the motor regions firing in an uncontrolled manner. After less than a minute the seizure subsides, though the person remains unconscious. This 'reset and start afresh' phenomenon is presumed to force the brain to restart, forgetting recent patterns of activity related to psychotic or neurotic symptoms. ECT is still widely used, primarily in cases of severe psychotic depression where imminent suicide is an acute risk, but the specific mechanism by which ECT produces effects is not known.

The immediate effect of ECT is usually to reduce the severity of symptoms to the point where recovery can occur, or where less severe therapies have a chance to enable the person to function independently. However, memory loss is very common in people who have received ECT.

The power of popular paradigms

The influence of Delgado and other high-profile charlatans is related to the unquestioning willingness of the public (and many scientists) to welcome a spectacular new idea, just because it fits in with current formulations (paradigms). The lack of real evidence does not seem to be an obstacle when the 'story' is attractive. The Piltdown hoax is an outstanding example of the power of a popular idea to blind science for decades.

Brighter prospects–Transcranial magnetic stimulation

Transcranial magnetic stimulation (TMS) was originally developed as a neurological tool to stimulate the motor pathways in a non-invasive manner. Stimulation of certain regions of the prefrontal cortex has been used in the treatment of depression. The resulting reduction in depressive symptoms is not as spectacular as for ECT, but is nonetheless highly significant. Even more importantly, TMS does not appear to cause any physical damage to brain, presumably because it involves only local stimulation at low threshold currents. Patients remain conscious and reasonably comfortable, and no anesthetic or muscle relaxant is necessary. A recent, stringently controlled study by Mark George has demonstrated significant reduction of symptoms in medication-resistant depressed patients treated with three weeks of daily repetitive TMS to the left prefrontal regions.

Magical magnetism

Unfortunately the magnetic component of TMS has suggested a link with the countless 'magnetic health' products which have been shamelessly marketed for a century or more. In TMS, a pulsed magnetic field well beyond the power of ordinary magnets is used to stir up electrical currents in brain tissue, causing depolarization and action potentials in neurons, and disrupting existing activity. Magnetic bracelets, wraps, blankets and belts all contain static magnets that do not pulse. There is no evidence that static magnetic fields affect neural firing. If they did, an MRI scan would induce a seizure or unconsciousness.

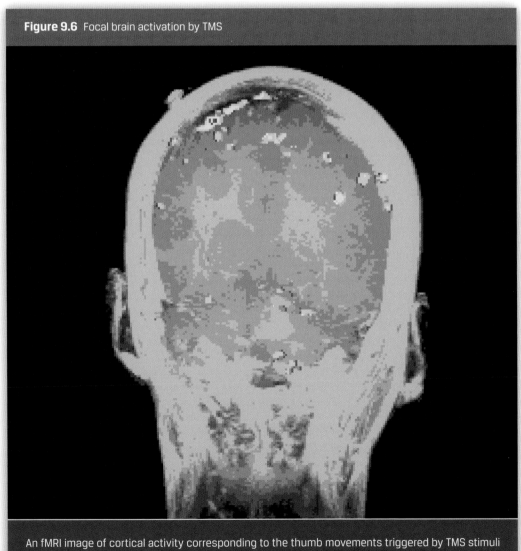

Figure 9.6 Focal brain activation by TMS

An fMRI image of cortical activity corresponding to the thumb movements triggered by TMS stimuli at the scalp location indicated. TMS permits electrical stimulation of small and precise cortical regions close to the skull. Image courtesy Dr Mark George, Medical University of South Carolina.

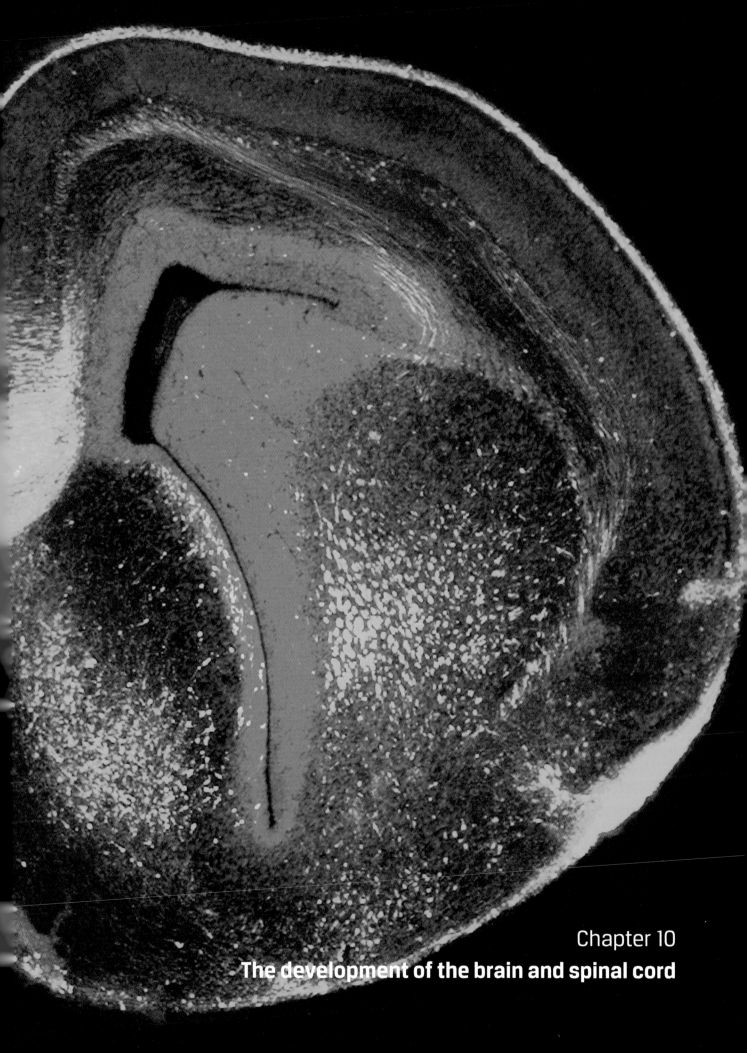

Chapter 10
The development of the brain and spinal cord

10 The development of the brain and spinal cord

In this chapter:

Genes and brain development	142
Early development of the brain and spinal cord	142
Regional development of the nervous system–segmentation and organizing centers	145
Formation of synapses	148
Axon guidance	149
Environmental influences on gene expression	150
Critical periods	150
Neural Plasticity	151
Later processes that refine the structure of the brain	152

Genes and brain development

The development of the brain is determined by the action of a large number of genes. Mammals have about 25,000 genes, and about half of these are primarily devoted to building and maintaining the brain. Since the adult human brain is estimated to have 86 billion neurons, each with thousands of connections, it is obvious that the genes provide general recipes, not specific blueprints, for building the brain. Because there are not enough genes to give specific instructions to individual neurons, the genes first generate basic groups of neurons and create chemical gradients to organize specialized connection patterns in each region.

Early development of the brain and spinal cord

The brain develops from a hollow cylinder of cells called the neural tube. This cylindrical structure forms the core of what will become the adult brain and spinal cord. At the top end of the tube, three swellings appear, the forebrain, midbrain, and hindbrain vesicles. A pair of additional vesicles grow out of the forebrain vesicle. These are the primitive cerebral hemispheres. The cerebral hemispheres will develop into an outer layer or cortex, and a collection of deep or buried cell groups which are commonly called the basal ganglia. The part of cortex to first form is the hippocampus, followed by the olfactory areas, and later the cerebral neocortex.

From the roof of the midbrain vesicle two pairs of cell groups bulge out. They are collectively referred to as the tectum, but are more commonly called the superior and inferior colliculi.

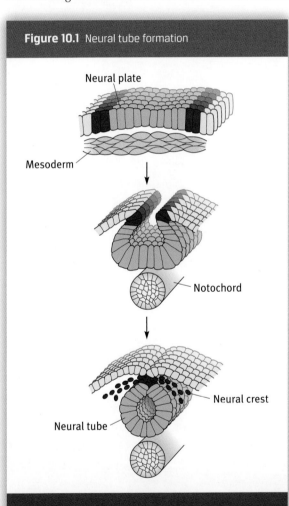

Figure 10.1 Neural tube formation

This series of diagrams shows the formation of the neural tube from the embryonic neural plate. Under the influence of secreted factors from the notochord, the cells of the neural plate sink down to form a groove and the edges of the groove (red) join to form the neural tube. The edges of the groove form the cells of the neural crest, which migrate away from the neural tube to form ganglia, melanocytes, and the skeleton of the face. (Diagram from Sanes, 2004, Figure 1.09, courtesy of Dr Dan Sanes, New York University).

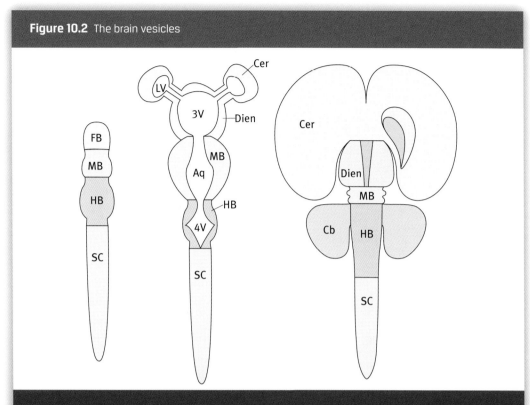

Figure 10.2 The brain vesicles

At the rostral end of the neural tube, three expansions appear-the forebrain, midbrain and hindbrain vesicles. The forebrain vesicle forms the cerebral hemispheres, the diencephalon, the hypothalamus, and the eye stalk. The hindbrain vesicle forms the isthmus, the rhombencephalon, and the cerebellum.

In the roof of the hindbrain, a diamond-shaped area forms the primitive fourth ventricle. In the border of the fourth ventricle adjoining the midbrain, a small outgrowth marks the position of the future cerebellum.

Neural crest

Just after the closure of the neural tube, a group of cells splits off from the dorsal surface of the neural tube and migrate away from the midline. This transient group of cells is called the neural crest. There is a line of neural crest cells on either side of the developing neural tube in the early embryo. The neural crest cells give rise to the ganglion cells of the autonomic nervous system and the sensory dorsal root ganglion cells of all spinal nerves. They also produce derivatives that are not directly related to the nervous system, the melanocytes of the skin and the bones and muscles of the face. The melanocytes migrate long distances to colonize all areas of the skin surface.

Multiplication of cells in the brain

The small group of embryonic cells that forms the neural tube quickly multiply and specialize to form specific functional centers in the brain. The rate of multiplication is extraordinary and the few hundred cells in the primitive neural tube eventually generate nearly 90 billion nerve cells in the adult brain. During the first two years of life there are times when up to 250,000 new cells are generated every minute.

Pasko Rakic and cortical development

Professor Pasko Rakic of Yale is a leader in research on cortical development. He demonstrated that newly formed neurons in the base of the cortex migrate radially to their final position. He has questioned the importance of the tangential migratory stream in primates.

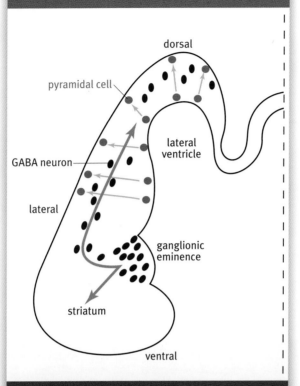

Figure 10.3 Radial and tangential migration in the cerebral cortex

Stem cells on the surface of the lateral ventricle develop and migrate toward the surface of the cortex to form the pyramidal cells. The cells involved in this radial migration are colored red in the diagram. A second population of migrating cells is formed in the ganglionic eminence of the basal forebrain. These are GABAergic cells which migrate tangentially through the cortex (purple cells along the pink arrow). These cells eventually form the population of inhibitory interneurons in the cortex. The cells of the ganglionic eminence also give rise to the GABAergic neurons of the striatum (caudate-putamen).

Migration of neurons

In many parts of the brain, the neurons continue to develop in the place where they were first formed. These neurons may send their axons to other places, but the neuronal cell bodies do not move. However, in a few places in the brain, groups of neurons migrate considerable distances to reach their final location. In the cerebral cortex, most of the cells are generated at the surface of the ventricle and migrate outward to find the cortical layer in which they will finally reside. Most of the neurons that migrate toward the surface of the brain (radial migration) are the future pyramidal cells of the cortex. On the other hand, the small interneurons of the cortex migrate tangentially from the base of the forebrain to reach their final settlement in the cortex. This remarkable tangential migration has been clearly demonstrated in rodents, but there is still debate about its significance in primates.

Another major source of migrating cells is the region that borders the roof of the hindbrain. This region, called the rhombic lip, gives rise to cells that migrate dorsally into the developing cerebellum, and others that migrate ventrally into the hindbrain. Because the tiny granule cells of the cerebellum eventually make up 70% of all the neurons in the brain, the rhombic lip can be said to be the most active cell generation area in the developing nervous system.

Creation of different functional groups of neurons–the proneural genes

In the very early embryo, some cells have the capacity to form either epidermal (body covering) cells or primitive neurons. The genes that guide this decision toward a neuronal fate are known as proneural genes. The proneural genes are represented in the embryo by ten genes called the basic helix-loop-helix (bHLH) genes. These bHLH genes make the transcription factors (the proneural proteins) that prompt neuronal development. It is difficult to understand how only ten proneural genes can prompt the formation of hundreds of different neuronal clusters. It is assumed that the proneural proteins interact with additional local factors to produce different varieties of neurons, but the evidence for this process is incomplete. Some well-known proneural bHLH genes are *Math1* (more recently called *Atoh1*), *Mash1*, *Ngn*, and *Tcf4*. The function of the bHLH proneural genes is closely linked to the genes of the Notch/Delta signaling pathway. For example, the rise in expression of *Ngn* in a differentiating neuron will inhibit neighboring uncommitted cells from developing into neurons. The inhibition is mediated by the Notch/Delta genes.

Regional development of the nervous system–segmentation and organizing centers

During the early development of the brain, the expression of a number of genes marks out the boundaries of different regions within a larger area. In the forebrain, the future cortex (pallium) is clearly marked off from the subpallial areas. In the diencephalon, three regions from rostral to caudal have been defined. They are referred to as prosomeres of the diencephalon. In the hindbrain a number of segmental compartments are created. Regional specification is determined by the expression of a wide range of genes. Here we will look at one example, the segmentation of the hindbrain.

Regional development in the hindbrain

The developing hindbrain is broken up into 12 regions from rostral to caudal. These regions are called the isthmus (the most rostral) and the rhombomeres. It was originally thought that there were only seven rhombomeric segments, but it is now clear that the total number is eleven. The

What is a brain segment?

Each of the segments of the diencephalon and rhombencephalon has a distinctive genetic signature. The cells in each segment show 'lineage restriction', a strong tendency to remain in the segment in which they were born, and not to migrate to neighboring regions.

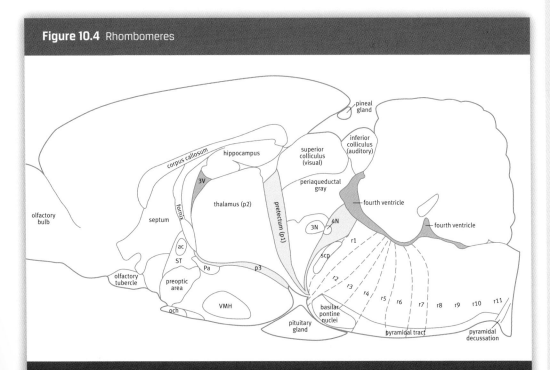

Figure 10.4 Rhombomeres

This is a diagram of an adult rodent brain showing the position of the segments of the hindbrain. The embryonic hindbrain contains twelve compartments from rostral to caudal. The first compartment, adjacent to the midbrain, is called the isthmus. The remaining eleven compartments are called rhombomeres. Rhombomeres 2 to 11 are defined by the expression of different hox genes. The isthmus is defined by the expression of fgf8. The cerebellum is formed by extensions of the isthmus and the first rhombomere.

Figure 10.5 Segmentation of the brain in a mouse embryo

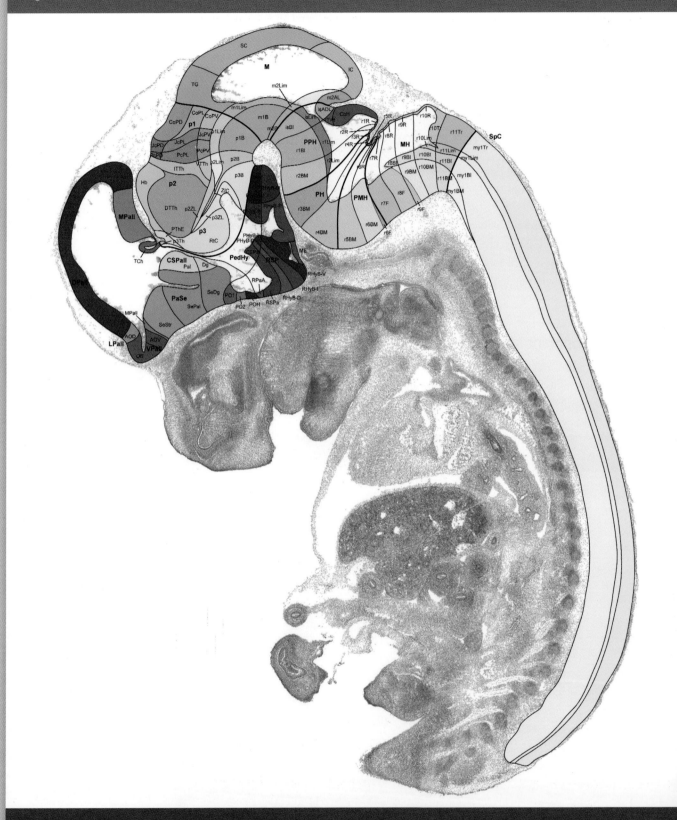

This is a photograph of a sagittal section of a mouse embryo (E13.5) with an overlay showing the subdivisions of the developing nervous system. The image is taken from the reference atlas developed by Luis Puelles as part of the Allen Developing Mouse Brain Atlas created by the Allen Institute for Brain Science. The citation for the image is Allen Developing Mouse Brain Atlas (Internet) Seattle (WA). Allen Institute for Brain Science ©2009. Available from: http://developingmouse.brain-map.org

genes responsible for creating the rhombomeres are the homeobox (hox) genes. The hox genes, which were originally discovered in developing insects, have been shown to be universally important for nervous system and musculoskeletal development in vertebrates. There are thirteen hox genes (*hox1-hox13*) in insects, but each gene has been multiplied four times in vertebrates. The four paralogs of each hox gene are named a, b, c, and d (e.g. *hoxa1, hoxb1, hoxc1, hoxd1*). The first four hox groups (*hox1-hox4*) are expressed in the hindbrain, and *hox5-hox13* are expressed in spinal cord, trunk, limbs, and tail. The isthmus and the first rhombomere are not directly controlled by the hox genes. Another factor that distinguishes the isthmus and first rhombomere from the rest of the hindbrain is that they alone give rise to the cerebellum.

In invertebrates, the role of the hox genes is to produce a segmented body– either a limited number of segments as in an ant or a lobster, or a very large number of segments as in a centipede. The basic mechanism of segmentation is the production of multiple versions of a single segment plan. In lobsters, for example, the segments are similar, but have different specialized projections (antennae, claws, or legs). The same segmental system rules the development of the hindbrain and face in mammals. The hox genes are responsible for the creation of the rhombomeric segments in the hindbrain, and also guide the production of facial bones and muscles associated with each hindbrain segment.

Organizing centers

At the junction of the developing midbrain and hindbrain, a region known as the isthmic organizer has powerful effects on the rostrocaudal development of the brainstem. A number of genes, including *wnt1, en1*, and *fgf8* have a strong influence over the further development of the midbrain and hindbrain (including the cerebellum). The importance of the isthmic organizer was first demonstrated by grafting experiments in developing birds and has been confirmed in mammalian studies. The isthmic organizer is one of a small number of organizing centers in the

Figure 10.6 The role of sonic hedgehog (Shh) as an organizer in the developing neural tube

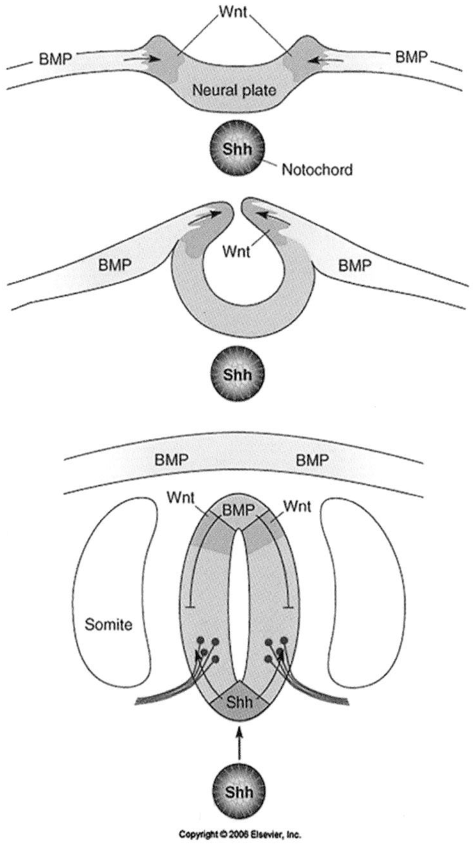

As the neural tube forms from the neural plate, its development is influenced by the expression of a gene called sonic hedgehog (Shh). Expression of sonic hedgehog in the notochord induces the development of ventral structures (such as motor neurons) in the neural tube. The dorsoventral patterning of the neural tube is created by the opposing influences of sonic hedgehog (ventrally) and BMP and Wnt genes (dorsally). (Diagram from Sanes et al, 2004, courtesy of Dr Dan Sanes, New York University)

The sonic hedgehog gene

Sonic hedgehog plays an important role in the early dorsoventral patterning of the neural tube. It is expressed in the basal plate of the neural tube, and induces the formation of motor neurons. The gene also has a number of other functions in the developing embryo. The hedgehog gene family was so named because of the spiky appearance of hedgehog mutants in fruit flies, and sonic hedgehog was named after a character in a video game.

developing nervous system. In each case, the organizing structure influences the development of nearby structures. In the development of the spinal cord, the notochord (which secretes *shh* [sonic hedgehog] gene) acts as an organizing center for dorsoventral patterning of the spinal cord.

Formation of synapses

Even while multiplication of cells is still going on, some of the neurons begin to sprout processes, which connect with other nerve cells or with muscle cells. The connections are called synapses. Much of the basic wiring of the brainstem and spinal cord is laid down early in fetal development. The most important early connections in the brain are those required for survival, such as those that control breathing and swallowing. There is no room for error here and these vital early connections are formed according to a strict program. The first movements of the fetus early in pregnancy in humans are clear evidence of the formation of early connections between neurons and muscle cells. In mice the connections with muscles are made after only 12 days, but in humans it does not happen until about 12 weeks of fetal development.

How motor nerves find the right muscles

An outstanding example of axon guidance is the way that motor neurons in the spinal cord link up with the 'right' muscles in the limbs. The Jessell group in Columbia University has demonstrated that there is a pairing system in which motor neurons expressing a particular gene connect with muscles that express the same gene. The system is very accurate, with over 95% of motor neuron axons finding the correct muscle.

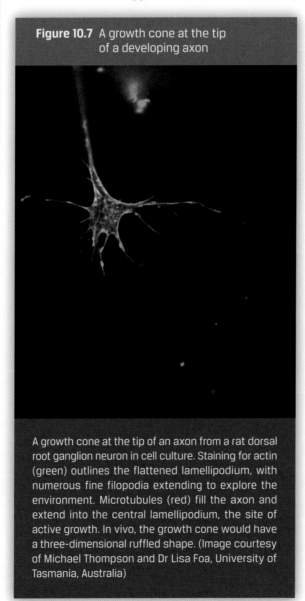

Figure 10.7 A growth cone at the tip of a developing axon

A growth cone at the tip of an axon from a rat dorsal root ganglion neuron in cell culture. Staining for actin (green) outlines the flattened lamellipodium, with numerous fine filopodia extending to explore the environment. Microtubules (red) fill the axon and extend into the central lamellipodium, the site of active growth. In vivo, the growth cone would have a three-dimensional ruffled shape. (Image courtesy of Michael Thompson and Dr Lisa Foa, University of Tasmania, Australia)

A second category of connections starts late in fetal development and extends into childhood and adult life. These later connections are not rigidly programmed, and they are made while the newborn child or animal interacts with its environment and is learning new skills. It was formerly believed that the formation of new connections stopped sometime soon after birth, but is now known that new connections are a normal part of the growth of all brains throughout life. The ability of the brain to constantly form new connections and to reshape patterns of connection in response to new information is referred to as plasticity. The plasticity of the human brain is remarkable. It has been estimated that the brain forms an average of a million new connections per minute throughout the whole of life.

Axon guidance

An axon that grows out from a parent neuron will only be successful if it connects with the right target cell. The mechanism that helps it find its correct target is called axon guidance, or axon pathfinding. In many of the examples that have been studied, growing axons have been found to be extraordinarily accurate in finding the correct path to their target. In some cases it has been shown that neurons find their correct partner because they have a common gene marker. For example, motor neuron clusters in the spinal cord and the muscle group to which they connect share a specific gene marker. It is presumed that the growing axon at first follows a general pathway but upon arrival it can detect the faint chemical signal from its target. A gene pairing system like this also exists in the hindbrain neuron groups that connect to particular cells in the cerebellum. Whatever the mechanisms, the formation of early connections is remarkably accurate and reliable, so that brains are always initially wired up in the same way.

At the tip of a growing axon is a very active enlarged area called a growth cone. The growth cone is a highly mobile structure that appears to feel its way through tissue, seeking very faint chemical signals that lead it towards its targets, or which steer it away from unwanted connections. The signals may be fixed in place, like signposts, or they may be distributed over long distances, like the breadcrumbs left behind by Hansel and Gretel. At the stage when neurons are maturing and sending out axons to make contact with other neurons, the developing nervous system secretes a tapestry of guidance molecules which attract and repel, defining pathways, marking crossing points, and identifying destinations. Growing axons explore this space with their growth cones, responding to some cues and ignoring others, according to the receptors they carry. Axons from diverse regions converge on shared pathways, then branch out to distinct targets as they encounter individual cues. A number of guidance molecules have been identified and their functions have been analyzed in detail, including netrins, semaphorins, and Slit proteins. Once successful in reaching their targets, axons stop growing when they encounter cell surface molecules that turn off growth and stimulate synaptic formation.

These 'stop signals' vary in concentration, so that arriving axons will fan out across the target tissue in an organized gradient. In the case of ephrins and Eph receptors, some axons will have few Eph receptors and some will have many. If they encounter a tissue with a gradient of ephrins expressed, axons with many receptors will stop in response to small concentrations of ephrins, whereas those with few receptors will push past low-ephrin areas and stop when the ephrins become more concentrated. Since the Eph receptors and ephrins are regulated by genetic gradients in the target and the origin of the axons respectively, this simple mechanism preserves topographic arrangements across long distances.

Synapse pruning and nerve cell death

Not all of these new connections work as well as they should, and millions of synapses are 'pruned' during early brain development so that the systems formed are accurate and efficient. Pruning of synapses continues to occur throughout life. The same 'pruning' process is applied to neurons, and many of neurons are killed off during development. There is evidence that neurons that have made poor connections, or that are unnecessary, are deleted. These waves of synaptic pruning and neuronal death are part of the process of building an efficient nervous system. As with synapses, the pruning of cells and axon branches continues throughout life.

Use it or lose it

The idea of non-functional cells dying off during neuronal development was long assumed to be true, but research in the laboratory of Professor Nobutaka Hirokawa has recently identified a possible mechanism. The axon transport protein KIF4 can keep neurons alive while they are active, by suppressing a regulatory enzyme. If the neurons remain unstimulated, the unsuppressed pathways act to shut the cell down and disassemble its proteins, a process called apoptosis.

Environmental influences on gene expression

Up to the time of birth, the development of the brain is driven by sets of genes. The genes are able to orchestrate the formation of the brain with its extraordinary wiring diagram, which enables the newborn animal to survive.

After birth, the development of the connections in the brain is controlled by a complex interaction between the organism and the immediate environment. The brain is continually shaped to adapt to the world around it. In turn, the expression of many genes is influenced by the environment. For example, continued severe stress raises cortisol hormone levels, and this may stop or alter the expression of some genes.

Critical periods

The concept of the critical period has received a great deal of attention by those interested in child development. More than sixty years ago, it was shown that kittens that had one eye kept closed during a particular period after birth would never develop binocular vision. The idea that there was a circumscribed period during which a particular input must be received in order for a part of the brain to develop had a major impact on theories of postnatal development. It also created a serious concern that children deprived of a particular stimulus or environment for a short period in their life would be deprived forever of some vital capability.

However, this view is not supported by contemporary evidence, and it is now felt that the concept should be applied more loosely. The new version of this concept accepts that critical periods are the ideal time for most children to receive a particular input, but argues that delayed inputs can still be effective.

The significance of critical periods is not yet fully understood. While they give a guide to the importance of the environment at particular stages, it would be a mistake to think that the critical periods are so rigidly specified that nothing can be done after the 'official' critical period has passed. For this reason, the term 'sensitive periods' is also used to denote longer time intervals during which experience can shape the nervous system.

Sensitive periods of human postnatal development that have received particular attention are those for language acquisition (mainly nine months to five years), development of emotional control (from birth to two years), motor control (from birth to eight years), and habitual ways of responding, such as enthusiasm and persistence (from six months to two years). However, in each of these examples, one must distinguish between the value of recognition of the sensitive period as an optimal time for input, versus the mistaken idea that it is a vital opportunity which, if missed, will be lost forever. The value of recognizing the opportunity in a defined sensitive period is important for policies for the optimal timing of interventions in preschool and primary school settings.

Critical periods in fetal development

Certain developmental insults can cause devastating damage to the developing brain and spinal cord. These include vitamin deficiency (neural tube defect), alcohol (fetal alcohol syndrome), and infections (rubella) which can cause widespread damage to the fetal brain. It is possible that cerebral palsy may also be caused by viral infections during pregnancy.

Neural tube defect (NTD), also called spina bifida, is caused by a disruption of the very early development of the neural tube, with incomplete closure in certain regions. The most common

---

Critical period for attachment

The opportunity to attach to a caring adult is crucial to the healthy development of human babies. The caring adult is normally the mother, but this is not essential if someone else is able to play the role. The critical period for attachment extends over the first two years, but the first year is the most important. Aggressive behavior in teenagers and adults has been linked to failure of attachment in childhood.

place for incomplete closure is in the caudal part of the spinal cord. Because the neural tube does not fully close, the spinal cord is disorganized, and the vertebral arch does not grow to cover and protect the spinal cord. When the baby is born, the membranes covering the spinal cord bulge out of the lower back. Surgery is required to construct a covering for the spinal cord, and in the most serious cases the disorganized spinal cord results in paralysis of the lower limb and loss of control of the bladder and bowel.

Some families have a genetic predisposition to NTD, but the most important factor is lack of the vitamin folate. If folate levels in the mother are low, the risk of NTD is very high. All women who might possibly become pregnant should make sure that their folate intake is high–either by eating leafy vegetables and other folate-rich foods, or by taking folate vitamin tablets as a supplement. In many countries, bread flour and breakfast cereals are enriched with folate to ensure that the whole population has adequate folate levels.

Rubella (German measles) infections during the first three months of pregnancy can cause major brain abnormalities, including deafness and blindness. Similar critical periods apply to the availability of nutritional elements, such as protein and fat. It is not well known that iodine deficiency during fetal development is the most important cause of mental retardation in the world–affecting millions of children in Asia and South-East Asia, and causing damage that will last for the rest of their lives. The problem of iodine deficiency is easily remedied with the supply of iodine supplements in salt and oil, but the logistic barriers in some areas are considerable.

Iodine deficiency and brain development

Iodine is an essential component of thyroid hormones. Iodine deficiency results in hypothyroidism, a failure to manufacture enough thyroid hormone. The thyroid gland becomes enlarged in a vain attempt to compensate. This kind of thyroid enlargement is called goiter. Thyroid deficiency during pregancy results in physical and brain growth retardation in the baby. This condition is called cretinism.

Neural Plasticity

The basic meaning of 'plastic' is something whose shape can be molded or changed, like clay. Brain plasticity, or neural plasticity, describes the ability of the mature brain to change itself, by generating new neurons or by making new connections. The realization that the brain can continually change itself is quite new, and has caused a significant change in neuroscience thinking in recent years.

For many years it was thought that the structure of the mature mammalian brain could not be changed, so that someone who suffered a stroke or some other form of brain injury might be condemned to some permanent loss of function. However, in the past thirty years it has become evident that the brain can compensate by creating major changes in connections after injury, and many rehabilitation programs have been developed to take advantage of this capacity.

The ability to grow and make new connections also occurs in the normal brain when it is asked to cope with new information. For example, an MRI study of the brains of final-year medical students found that the gray matter in their parietal cortex and hippocampus increased substantially in the period leading up to their final exams.

Some parts of the mature mammalian brain, notably the hippocampus, the cerebellum, and the olfactory bulb, can actually generate new neurons. In most other areas of the brain, it seems neuroplasticity is solely the result of massive reorganization of synaptic networks in response to new experiences.

A pioneer in the area of neuroplasticity, Mike Merzenich, has promoted the idea that exercising the brain can increase its capacity as a defense against cognitive deterioration in elderly individuals.

Later processes that refine the structure of the brain

The development of the brain does not stop in childhood. As noted above, the formation of new synapses continues throughout life. In the cerebral cortex, this is accompanied by continued growth of dendrites. It has been shown that there is a marked increase in new synapse growth in the frontal lobes during adolescence, so that the frontal lobes are not fully mature until about 25 years of age.

Another vital process that occurs during childhood and adolescence is that of progressive myelination. Myelin sheaths are laid down during the postnatal years until early adulthood. Myelination increases the speed of transmission of action potentials along axons, and might provide a mechanism for stabilizing connection times while the body is enlarging. The distance from spinal cord to the foot muscles increases from about 30 cm in a baby to about 1 m in an adult; if myelination did not occur it could take thirty times as long for a command to reach the muscles in an adult.

The development of the cerebellum is delayed compared with other parts of the brain. At the time of birth, the cerebellum is extremely small, but during the first six years of life in humans it grows to the size of an orange. The role of the cerebellum is to assist with the coordination of movement, so its development is delayed until the period when the child is learning new motor skills.

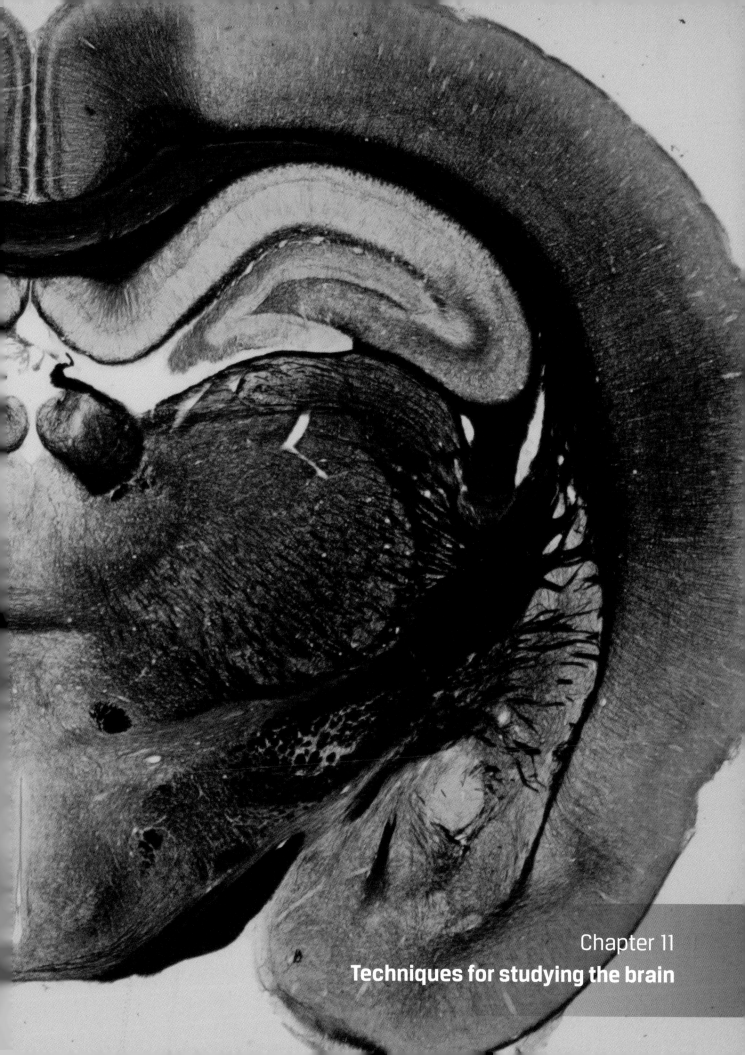

Chapter 11
Techniques for studying the brain

11 Techniques for studying the brain

In this chapter:

Cutting thin sections of the brain	154
Staining brain sections	154
Cell culture	160
Hodology: using tracers to map connections	160
Molecular genetics	161
Non-invasive imaging techniques	162
Functional imaging	163
Electrophysiology	164

Most of the information we have about the brain has come from the study of thin sections that have been stained to show nerve cells and their processes. The most famous of these studies are those conducted by a Spanish neuroanatomist, Ramón y Cajal. Cajal used a special silver stain to study the main neuron types found in the brain. During the second half of the twentieth century, a range of techniques were developed that enable us to trace connections between different parts of the brain. In recent years, a new generation of imaging techniques has allowed us to look at brain anatomy in live subjects.

Cutting thin sections of the brain

Microscopy requires tissue to be sliced (sectioned) very thinly. The brain is usually prepared for sectioning by treatment with a chemical preservative such as formalin, but fresh tissue can be sectioned using a microtome with a vibrating blade, or by cutting sections of brain that has been frozen. Thin sections are cut on a microtome (a machine that can cut sections at thicknesses between 5 and 100 micrometers). When fixed tissue is prepared for sectioning, it is usually dehydrated in alcohol and then embedded in paraffin wax or celloidin. Each of these methods presents specific challenges: fixation can permanently change some of the molecules in a neuron, whereas frozen sections are more difficult to handle and may become torn.

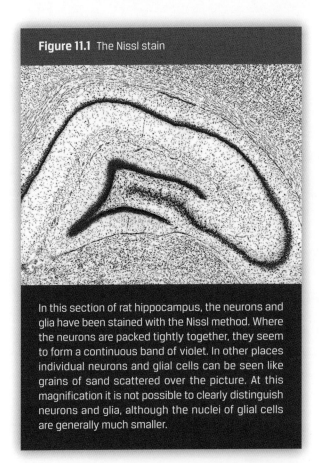

Figure 11.1 The Nissl stain

In this section of rat hippocampus, the neurons and glia have been stained with the Nissl method. Where the neurons are packed tightly together, they seem to form a continuous band of violet. In other places individual neurons and glial cells can be seen like grains of sand scattered over the picture. At this magnification it is not possible to clearly distinguish neurons and glia, although the nuclei of glial cells are generally much smaller.

Staining brain sections

The routine ways to stain brain sections are with a stain for neuronal nuclei and RNA (most commonly the Nissl stain) and a stain for myelinated axons (such as the Weigert stain or Luxol fast blue). However, more information can be gained by using histochemical stains and tracers for neuronal connections.

Nissl stains use a variety of dyes (e.g. thionin, cresyl violet, fluorescent compounds) to show charged structures (Nissl bodies) in the soma of neurons and glia. The Nissl stain is most intense in nucleoli and in the rough endoplasmic reticulum of neurons. For myelinated axons in the nervous system, a variety of techniques selectively label the unique physical properties of the densely wound membranes. Some preparation methods deposit silver haematoxylin in the protein scaffold of

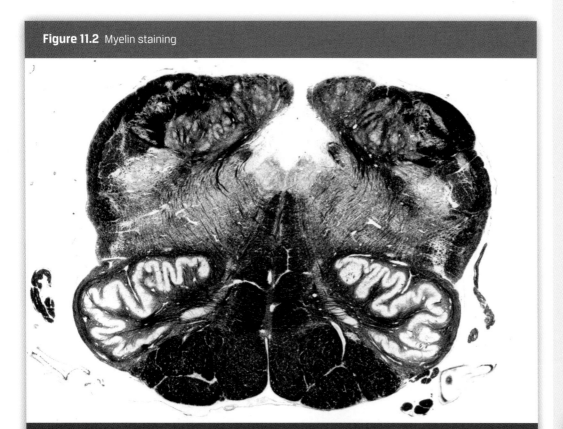

Figure 11.2 Myelin staining

In this section of human hindbrain the myelin has been stained with the Weigert method. The stained myelin appears brown or black. At the bottom of the section there is dense brown staining in the corticospinal tracts.

the membranes. Other myelin stains use dyes such as luxol fast blue or osmium salts, which stain the fatty content of the sheaths.

Histochemical staining

Histochemical stains are used to mark particular parts of cells in the brain based on physical and chemical properties. Typically they involve chemical reactions which render specific cellular components into particular chemical states, followed by application of dyes attracted to those properties, or reactions in which colored products are deposited on specific types of structures. Histochemical techniques for detecting enzymes usually work best on unfixed or lightly fixed sections, whereas some of the more chemically aggressive methods require robustly preserved material.

The Golgi silver impregnation technique relies on chemical preparation of thick blocks of tissue, after which grains of metallic silver crystallize inside the membranes of individual cells, producing a dense black precipitate that highlights every detail of the cell body and dendrites against a golden background. If this occurred in every cell the block would be solid black, but for unknown reasons only about 1% of neurons are stained by Golgi methods. This allows extraordinary detail to be seen in a tiny subset of cells, and enabled Cajal and other investigators of neuroanatomy to view and infer the connectivity of the nervous system in amazing detail. Frequently, only the cell body and dendrites are labeled by Golgi deposits, and myelinated axons are not stained.

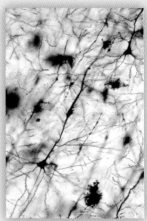

Figure 11.3
Golgi staining of cortical cells

This is a thick section of cerebral cortex stained with the Golgi method. This method stains only about 1% of the cells, so the black cells and processes can be seen against a pale background. In this section a number of pyramidal cells can be seen. They have triangular cell bodies and long, thick apical dendrites. (Image courtesy of Professor Sandra Rees, University of Melbourne).

NADPH-d (nicotinamide adenine dinucleotide phosphate-diaphorase) histochemistry employs a chemical reaction that deposits a vivid blue dye in the presence of enzymes which synthesize nitric oxide, providing a rapid indication of which cells use nitric oxide as a neurotransmitter.

Cytochrome oxidase (CO) histochemistry combines a chromagen (DAB) with a substrate molecule to indicate where this mitochondrial enzyme is active. In neural tissue the great majority of CO activity is found in inhibitory interneurons, so the density of staining can indicate long-term patterns of inhibitory activity in tissue.

Acetylcholinesterase (AChE) biochemistry uses s-acetylthiocholine iodide and ethopropazine to deposit a brown product wherever this breakdown enzyme is present and active. Since a large proportion of neurons have receptors for acetylcholine, this technique can be used to characterize regions of strong cholinergic activity and differentiate between different areas in the brain. AChE visualization has proved very valuable in the preparation of detailed atlases of the brain.

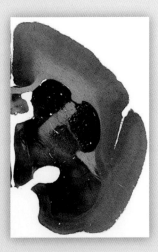

Figure 11.4
NADPH-d staining

This is a coronal section through one side of a marmoset forebrain. NADPH-d staining is strongest in the caudate and putamen (middle of the section) and in some parts of the amygdala (lower part of the section). (Image courtesy of Dr Hironobu Tokuno, Tokyo Metropolitan Institute for Neuroscience)

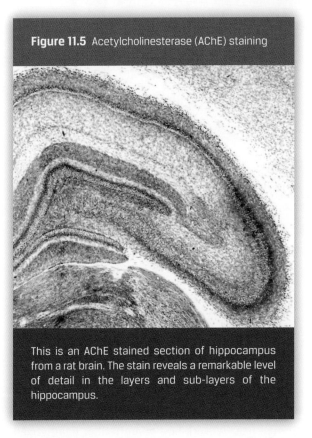

Figure 11.5 Acetylcholinesterase (AChE) staining

This is an AChE stained section of hippocampus from a rat brain. The stain reveals a remarkable level of detail in the layers and sub-layers of the hippocampus.

DNA can be visualised using dyes such as 4',6'-diamidino-2-phenylenindole dihydrochloride (DAPI) or Hoechst 33258, which form fluorescent complexes with the DNA molecule. These dyes show large neuronal nuclei as well as the small nuclei of glia. DNA being broken down as part of apoptosis is visualised by TUNEL (deoxynucleotidyltransferase-mediated dUTP nick-end labeling) stains, which indicate programmed cell death or DNA damage.

Calcium ions can be imaged by loading cells with fluorescent dyes (e.g. fura-2, indo-1) whose intensity varies according to calcium concentration. These techniques are often employed in live neurons in slices or in cell culture, enabling the regulation and traffic of this important ion to be tracked in real time.

Zinc and other metal stains use protocols that typically deposit silver (Timm-Danscher stain–Figure 11.6) or fluorescent dyes around ions of zinc and other heavy metals. Zinc is used in DNA regulation and synaptic plasticity as well as other cellular signaling, so zinc stains can show long-term patterns of activity and responses to change.

Immunohistochemistry

Immunohistochemical stains can detect the presence of particular proteins in neurons. The most common way to detect particular proteins is based on the specificity of immunoglobulin G (IgG) antibodies derived from common lab animals. The common types are polyclonal and monoclonal. Polyclonal Ig antibodies are derived from the serum of an animal inoculated with foreign proteins,

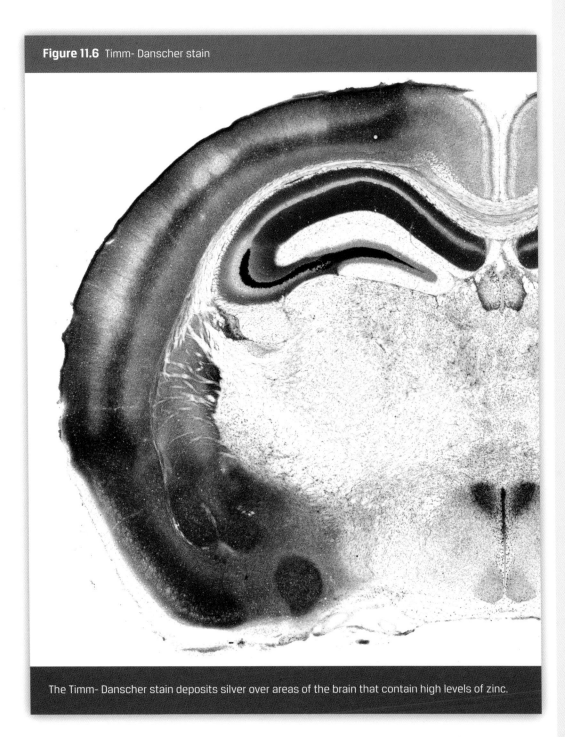

Figure 11.6 Timm-Danscher stain

The Timm-Danscher stain deposits silver over areas of the brain that contain high levels of zinc.

allowing the animal to generate a range of IgG antibodies to the foreign proteins. For monoclonal antibodies an antibody-producing cell is induced to grow in a clonal tumor mass. This mass secretes a fluid loaded with a single type of antibody, specific to one site on the protein.

Once these primary antibodies are applied to tissue sections they bind to their specific protein, but they usually cannot be visualized without an additional step. This is done with the help of a secondary antibody, which is chemically modified to carry either a fluorescent molecule or a catalytic molecule which can be stained. The most common catalytic molecule is a peroxidase, that can then deposit a chromagen such as DAB (3,3'-diaminobenzidine) for permanent labeling. Fluorescent labels can be used in situations that require double-labeling to view multiple proteins simultaneously, and can be used in controlled laser excitation techniques such as confocal and two-photon microscopy for three-dimensional localization.

The technical challenges of immunohistochemistry include ensuring that antibodies can accurately detect the proteins of interest (for which a positive control is often employed) and that they do not detect other proteins. Antibodies depend on specific structural qualities of the proteins concerned, but many of the techniques used to fix and stabilize tissue sections can have drastic effects on protein structure. For example, aldehyde fixation strongly cross-links proteins, which may interfere with antibody binding.

Common immunohistochemical markers

Calcium binding proteins (such as calbindin, calretinin, and parvalbumin) are frequently labeled in studies of the central nervous system. These markers are often restricted to particular cell groups, and are very useful in anatomical mapping studies.

Immunohistochemical markers and brain mapping

The use of a panel of immunohistochemical markers has enhanced the ability of neuroscientist to map animal and human brains. Characterizating of a nucleus in the rat with a particular pattern of markers can assist with the identification of the homologous nucleus in the brains of humans and other species.

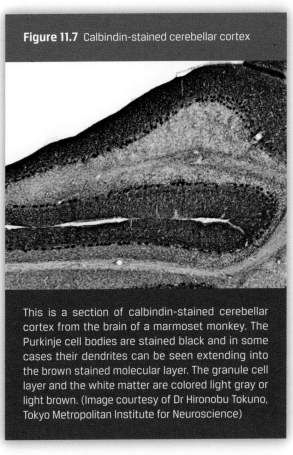

Figure 11.7 Calbindin-stained cerebellar cortex

This is a section of calbindin-stained cerebellar cortex from the brain of a marmoset monkey. The Purkinje cell bodies are stained black and in some cases their dendrites can be seen extending into the brown stained molecular layer. The granule cell layer and the white matter are colored light gray or light brown. (Image courtesy of Dr Hironobu Tokuno, Tokyo Metropolitan Institute for Neuroscience)

Neurofilaments are structural proteins with transport-related functions used by a proportion of neurons. The monoclonal antibody SMI-32 labels neurofilament triplet proteins in dephosphorylated states. The varying expression of these proteins across different structures of the nervous system make SMI-32 a popular label for investigating anatomy, although its functional significance is not yet clear.

Several glial markers are commonly used to distinguish specific glial cell types. GFAP (glial fibrillary acidic protein) identifies astrocytes, particularly in their 'reactive' state. Microglia uniquely contain the protein ferritin, and oligodendrocytes are detected by a range of markers such as myelin basic protein and CNP (2'3'-cyclic nucleotide-3'-phosphohydrolase). Polydendrocytes are specifically labeled with NG2, a proteoglycan molecule on the cell surface.

NeuN, an antibody derived from mice inoculated with purified neuronal nuclei, is specific to an unknown nuclear protein expressed by nearly all neurons and no glial cells. Unlike Nissl staining, this allows neurons to be visualized separately from glia. A few neuron types (such as the mitral cells of the olfactory bulb and the Purkinje cells of the cerebellum) do not express the NeuN protein and are therefore unlabeled.

BrDU labeling uses immunohistochemistry to detect an artificial nucleotide in proliferating cells. The technique depends on cells taking up an injected thymidine analog (bromodeoxyuridine, BrDU) during periods of DNA synthesis, and the detection of the BrDU analog by a monoclonal

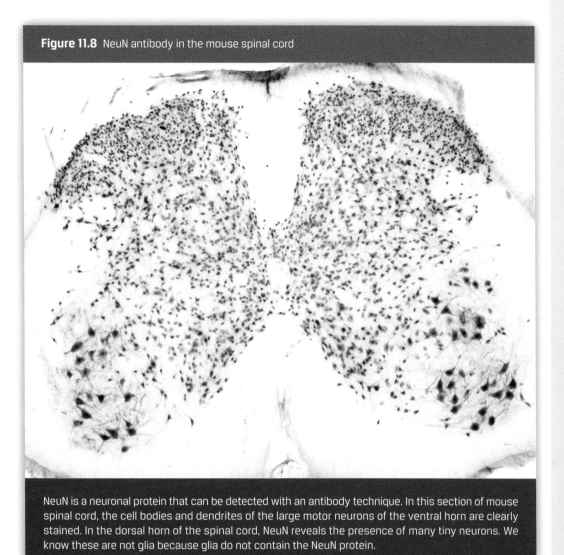

Figure 11.8 NeuN antibody in the mouse spinal cord

NeuN is a neuronal protein that can be detected with an antibody technique. In this section of mouse spinal cord, the cell bodies and dendrites of the large motor neurons of the ventral horn are clearly stained. In the dorsal horn of the spinal cord, NeuN reveals the presence of many tiny neurons. We know these are not glia because glia do not contain the NeuN protein.

Using NeuN to count neurons

The Brazilian neuroscientist Suzana Herculano-Houzel has developed a clever technique to count neurons in whole brains or in parts of brains. The tissue is homogenized with detergent in a blender in order to free all the nuclei from the cells. A small sample is withdrawn and stained with NeuN so that the neurons, and not the glia, can be accurately counted.

antibody. Since thymidine is not part of RNA, the only cells using thymidine are those creating new DNA as part of cell division or DNA repair. Accordingly BrDU labeling is often used in conjunction with markers of cellular immaturity as evidence for cell division, specifically the process of neurogenesis, in which new neurons are produced. A recent variant of this technique, using ethynyldeoxyuridine (EdU), directly incorporates a fluorescent nucleotide substitute into newly synthesized DNA.

Lectin stains

Lectins are plant proteins that bind extremely strongly to specific patterns of sugars in carbohydrate molecules. Like many cells, neurons exhibit glycoproteins on their cell membranes as part of their interactions with other cells. Applying biotin-conjugated lectins to tissue sections will label them according to glycoprotein distribution, allowing distinctions to be made on these criteria.

Cell culture

Another way of studying nervous system cells is to remove them from the living animal (the in vivo setting, meaning 'in life') and keep them alive and growing in an artificial setting (in vitro, meaning 'in glass'). To do this the cells must be kept in a culture medium, a solution of ions and nutrients designed to resemble the extracellular fluid environment in which the cells normally live. As long as the medium allows for gas exchange across its surface, or by bubbling gas mixes through it, cells can stay alive for weeks and even grow and differentiate.

Monocultures use isolated preparations of a single cell type, separated by a choice of tissue collected and the medium used to keep cells alive. This allows large groups of similar cells to respond to experimental conditions in large numbers. The weakness of this technique is that the cells are deprived of their normal interactions with other cells, and are therefore in a situation which is very different from their normal environment in vivo. Co-cultures use multiple cell types grown on the same surface or in close proximity to each other in order to better reproduce the interactions of cells. When the mix of cell types is the same as those found in the original tissue it is referred to as an organotypic culture. Slice culture refers to the technique of taking slices about 1 mm thick from chilled but still-living tissue, and maintaining them in a culture medium. In this way the relationships between cell types, and many of the interconnections between them found in vivo, are preserved intact, and by careful selection of the slice plane it may be possible to investigate quite complex connected systems, such as thalamo-cortical interactions. Removing any nervous tissue from its normal context in the organism will distort its function to a greater or lesser extent. Because of this, the usefulness of data derived from culture preparations depends both on the care taken to compensate for the change of environment, and on the choice of culture technique in order to minimize the effects on the system under study.

Hodology

This is the name for the study of connections in the nervous system. It comes from the Greek word 'odos' meaning a road or a pathway. Connections in the nervous system can be mapped with either retrograde or anterograde tracing methods, or with electrophysiology.

Hodology: using tracers to map connections

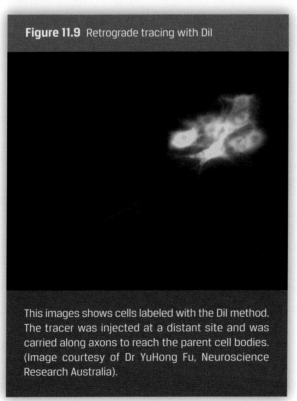

Figure 11.9 Retrograde tracing with DiI

This images shows cells labeled with the DiI method. The tracer was injected at a distant site and was carried along axons to reach the parent cell bodies. (Image courtesy of Dr YuHong Fu, Neuroscience Research Australia).

Most techniques for mapping the connectivity of the nervous system (called hodology) often rely on the transport mechanisms used by neurons to move proteins and organelles along their axons. Tracers can reveal the connections of neurons. They can be used to track the growth of axons, and can show regrowth after injury. Tracers may be delivered in solid form (a dye crystal or bead), by pressure injection, or by iontophoresis (in which a polar compound is infused into the tissue by passing an electric current through a pipette electrode containing the tracer), or via infectious agents (such as modified rabies virus or adenovirus). Tracers may be anterograde, that is traveling from cell body to axon terminals (e.g. cholera toxin B, dextran-biotin), or retrograde, that is, taken from axon terminals back to cell body (e.g. WGA-HRP, wheat germ agglutinin-horseradish peroxidase, fluorogold, fast blue). A few tracers are bi-directional (e.g. DiI and other dextran compounds). Most tracers are used in living tissue, but some (such as DiI) can be used to trace axons in postmortem tissue.

The value of tracing data depends on delivery, uptake, and visualization. Tracers have to be placed on the appropriate cells or terminals in order for the data to be useful. Selection of the delivery site is usually with the use of a stereotaxic atlas. Stereotaxic atlases define specific anatomical landmarks (e.g. bregma) that can be located during surgery, and the location of regions, nuclei and tracts are specified with reference to these landmarks. This allows tracers to be placed with great precision using a stereotaxic frame, a precise three-dimensional positioning apparatus. Some tracers can be detected in target areas in less than a day but others take weeks to reach their destination. Tracers can be detected with fluorescence techniques, or with histochemical techniques (such as chromogens to detect HRP). Fluorescent tracers may also be photoconverted using light-dense chromagens such as DAB to make a permanent label.

Molecular genetics

In-situ hybridization (ISH) uses tagged pieces of DNA that stick to complementary RNA molecules. Conventional ISH uses radioactive tags that make dark dots on photographic emulsion to show which cells are producing RNA for a particular gene. ISH provides an indication of which cells are actively translating a particular gene into a protein at the time of application, but they cannot reveal the previous history of translation.

The ability to target genes in mice was developed by 2007 Nobel Prize winners Capecchi, Evans, and Smithies. Gene targeting can be used to knock-out genes, or to insert (knock-in) new genes into developing cells. Gene knock-ins and knock-outs use techniques that can add a functioning gene to the organism, or disable a gene it already has, so that its effect on the animal can be tracked through development and adulthood. Refinements of this technique include inducible knock-ins, in which the new gene can be triggered to express at a particular time. A commonly used induction system relies on triggering of an estrogen receptor with the drug Tamoxifen. The animal that develops will also usually produce the reporter molecules whenever it translates the protein. Reporter molecules allow the pattern of expression to be viewed by histochemical visualization methods, or fluorescence microscopy.

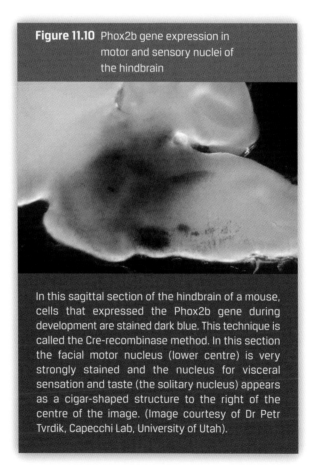

Figure 11.10 Phox2b gene expression in motor and sensory nuclei of the hindbrain

In this sagittal section of the hindbrain of a mouse, cells that expressed the Phox2b gene during development are stained dark blue. This technique is called the Cre-recombinase method. In this section the facial motor nucleus (lower centre) is very strongly stained and the nucleus for visceral sensation and taste (the solitary nucleus) appears as a cigar-shaped structure to the right of the centre of the image. (Image courtesy of Dr Petr Tvrdik, Capecchi Lab, University of Utah).

Fate mapping

The ability to map the migration and final destination of an embryonic cell or cell group has been the goal of embryologists for over a century. The chick-quail grafting technique developed by Professor Nicole Le Douarin in France gave this area new life in the 1970s and 1980s. However, the recent development of cre-recombinase lineages has provided a powerful tool that allows embryologists and anatomists to map the precise fate of cells which express a particular gene in the embryo.

Non-invasive imaging techniques

The most common images of the human nervous system are provided by non-invasive or minimally invasive scanning techniques that can be applied to living tissue. Normal X-ray images cannot reveal much about brain structure, but the use of contrast media injected into the bloodstream, or air introduced into the ventricular spaces can be used to enhance X-ray images of the brain. Unfortunately, these enhancement procedures are technically difficult and occasionally dangerous. A significant improvement in X-ray imaging came in the 1980s from computer reconstruction of multiple X-ray images from a moving source, using mathematical techniques to deduce the three-dimensional distribution of tissue. These computed tomography (CT) scans can show gross outlines of nervous system structures, and can detect alterations in density, caused by strokes, or some types of tumors.

The impact of MR scanning on neurological diagnosis

Until the late 1970s, investigation of brain disorders was limited to skull x-rays and injection of tracers into blood vessels. The introduction of MR scanning has had a huge impact on neurological diagnosis, because the detailed structure of the brain can be seen in living subjects. Because the scanning is not invasive, it can be repeated to follow changes in the brain.

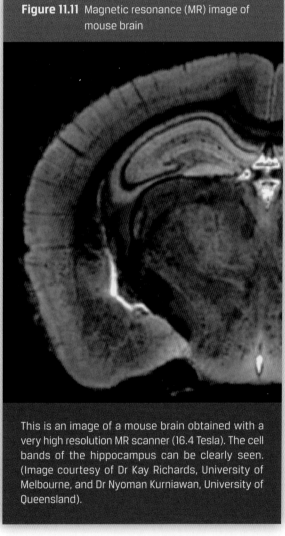

Figure 11.11 Magnetic resonance (MR) image of mouse brain

This is an image of a mouse brain obtained with a very high resolution MR scanner (16.4 Tesla). The cell bands of the hippocampus can be clearly seen. (Image courtesy of Dr Kay Richards, University of Melbourne, and Dr Nyoman Kurniawan, University of Queensland).

A major breakthrough in the 1990s resulted in the use magnetic resonance (MR) for imaging the brain. MR relies on the ability to excite specific types of atoms–usually hydrogen in water molecules–and to image them in a systematic way. This allows three-dimensional maps to be obtained. Since neuropil and fiber tracts vary in their hydrogen content, detailed delineation of nervous system structures can be achieved, with the resolution dependent on the strength of the magnetic field used. Early MR imaging used field strengths of 1-3 Tesla, yielding resolutions of 1-3 millimeters, but recent small-volume MR scanners use 4-16 Tesla fields, which can resolve structures in the 20-50 micrometer range.

Diffusion tensor imaging (DTI) uses MR scanners to obtain information about fiber orientation in the brain by structuring the scanning pulses to pick up how freely molecules are moving by diffusion along axons. Using three axes of diffusion scanning, every point in the brain is assigned an intensity value along with a mathematical description of the freedom of movement in three dimensions (called a tensor). The result is an image that gives an indication of which way axon bundles run in different parts of the brain.

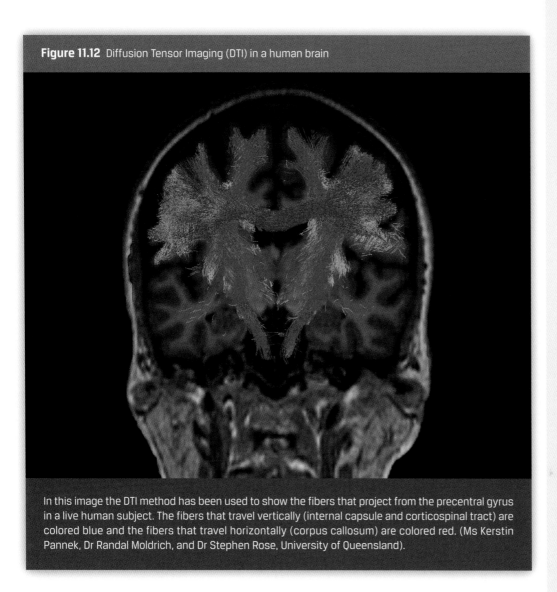

Figure 11.12 Diffusion Tensor Imaging (DTI) in a human brain

In this image the DTI method has been used to show the fibers that project from the precentral gyrus in a live human subject. The fibers that travel vertically (internal capsule and corticospinal tract) are colored blue and the fibers that travel horizontally (corpus callosum) are colored red. (Ms Kerstin Pannek, Dr Randal Moldrich, and Dr Stephen Rose, University of Queensland).

Functional imaging

Functional imaging methods depend on detecting changes in blood flow or metabolic consumption in active regions of the nervous system. Functional MRI (fMRI–also called Blood-Oxygen-Level-Dependent MRI -BOLD MRI) depends on differences in MRI signal intensity produced by oxygenated blood levels in tissue. A reference structural scan is made while the subject is resting. Additional scans are performed while the subject performs a task designed to access a particular ability or mental state, and differences in intensity are found by subtracting the control scan values. The differences are then color-coded and overlaid on the structural scan. Blood flow changes are mediated by astrocytes and depend on neuron-astrocyte interactions during periods of high activity. These changes are greatest in the cerebral cortex, which is why fMRI studies tend to show most significant changes in cortical regions. Although other parts of the nervous system may be just as involved in the task, they may not manipulate their blood flow as much and hence remain undetected. Another problem with fMRI is deciding whether the increased activity is due to active processing, suppression, or activity that is only peripherally related to the task.

Positron emission tomography (PET) and single photon emission computed tomography (SPECT) use radioactive substances administered intravenously. A three-dimensional distribution map is produced by scanning for a long period with a gamma ray detector. The radioactive markers used are designed to be taken up by active cells but are resistant to being broken down, therefore they remain in the cells and emit radiation for a brief period. As such this technique is not recommended for frequent or ongoing use in children.

Electrophysiology

Electrical recording and stimulation techniques have been used to study the nervous system for 200 years. They rely on the fact that the activity of neurons and glia is associated with changing electrical potentials, and that these electrical events can also be artificially provoked by applying electrical currents. In recent times, this field has tended to focus on understanding how the nervous system codes and processes sensory and motor control information in terms of action potentials.

Sensitive amplifiers are used to magnify the voltage changes caused by graded action potentials in receptors, primary afferent neurons and other regions of the nervous system. Electrodes may be constructed from glass pipettes, sharpened wires with insulated coatings, silver/silver chloride wires, or carbon fibers. The electrodes can be inserted inside cells, 'patched' onto the cell membrane, or positioned extracellularly or near groups of neurons to record field potentials. Amplified signals can be monitored (visually and audibly), digitized, timed, averaged, and classified.

Extracellular recordings use sharp, insulated electrodes to penetrate into neural tissue and measure electrical potentials for recording and analysis. Electrode design and construction influence the types of signals that can be detected. For example, the electrical impedance of the electrode dictates the size of the tissue volume in which it detects signals. A low-impedance electrode might collect data from tens or hundreds of neurons, whereas a high-impedance electrode might be able to isolate the responses of a single cell from the background of general activity. Recently, array electrodes with several (4-100) large-area, low-impedance electrodes have been used to sample activity across a large volume of tissue; comparison of the signals seen by each electrode allows spike sorting, which attributes each measured action potential to an individual cell, based on consistencies in which electrodes detect it most strongly. In this way the activity of dozens of individual neurons can be measured simultaneously.

The most refined type of electrophysiological investigation of cells is the patch clamp, in which a small area of membrane is sealed into a glass electrode tip and becomes part of an electrical circuit. Current balancing electronics permit an extraordinarily detailed investigation of ions passing through the membrane and the resultant changes in voltage. Patch clamps can be used on slice-cultured cells, or in living brains, so that their responses can be explored in a realistic setting. The glass electrode also allows different solutions to be applied to the membrane surface. Patch clamp amplification is so sensitive that the flickering open-close behavior of individual ion channels has been recorded at nanosecond time-scales.

A pioneer of electrophysiology

The Australian Sir John Eccles was a pioneer in the area of sophisticated electrophysiology. He recognized small excitatory and inhibitory potentials on motor neurons, and demonstrated the role of acetylcholine as a neurotransmitter. He was awarded the Nobel Prize in 1963 for his work on the physiology of synapses.

Sensory electrophysiology

Computer control of stimuli enables systematic analysis of the effect of sensory input on particular neurons. Electrophysiological experiments require careful controls, such as the systematic variation of stimulus parameters, and careful monitoring of anaesthesia and physiology in order to make the results accurate and reproducible. The electrical properties of the recording setup are also crucial, since the presence of noise can ruin the detection of responses. Shielding, earthing, signal processing, and digitization all need to be optimized.

Recording neuronal activity from the skin surface

The large-scale electrical activity of groups of excitable cells stirs up tiny electric currents in the tissue around them. These currents can be recorded and studied, using electrodes placed on the skin over the area of interest. Although the signal is a combination of millions of individual action potentials, the massed activity can take on different qualities under various circumstances, usually described in terms of frequency and amplitude. EEG (electroencephalography) is an example of this technique applied to recording brain activity.

By placing electrodes at multiple points on the scalp, EEG recordings can provide significant information about the spatial distribution, amplitude, timing and frequency composition of the electrical potentials from the brain. Because skin, bone, membranes and fluid tend to spread the signal out and reduce its size, EEG recordings depict large-scale electrical events in the cortex lasting from milliseconds to several hours. At these longer time scales and larger spatial scales, the mass action of the tissue is strongly influenced by astrocytes as well as neurons, and reflects changes in the overall activity and excitability of the cortex. These patterns shift dramatically through the course of sleeping and waking, as well as during epileptic seizures. Careful control of stimulus presentation and the use of computer averaging, allow components of the EEG to be extracted as event-related potentials (ERPs).

Magnetoencephalography (MEG) measures similar scales of activity but instead uses sensitive quantum devices to detect the minuscule magnetic fields given off by electrical events. MEG has somewhat lower resolution than EEG but can deduce the signal sources in three dimensions, providing a better visualization of the brain structures involved.

Appendix A
Voltages, potentials, and cell membranes

Membrane potential

A membrane potential is an electrical voltage that can be measured across the membrane of a cell. Membrane potential is usually expressed as the voltage of the inside of the membrane compared to the outside, in the same way that the 9 volts of a battery is the voltage difference between its two terminals. Voltages are called potentials because they have the potential to drive electric currents and do work.

The membrane potential in neurons is determined by differences in the concentration of sodium and potassium ions between the inside and outside of the cell. The differences in concentration are created by the ability of the membrane to tightly control the movement of ions into or out of the cell.

Neurons use metabolic energy to draw potassium ions into the cell and push sodium ions out. The result is that potassium is about thirty times more concentrated inside than outside, and sodium is about ten times more concentrated outside than inside.

There are three types of membrane potential–the equilibrium potential, the resting potential, and the action potential. The next section will deal with each of these in turn.

Equilibrium potential

Equilibrium potential refers to the membrane voltage that would balance inward and outward flow rates for one particular kind of ion. Every kind of ion (sodium, potassium, chloride, sulfate, calcium, etc.) has its own equilibrium potential, which depends only on its concentration inside and outside the cell. For example, potassium ions pass through membrane channels in both directions with relative ease, but the rate at which they leak out is much greater due to the concentration difference (high inside and low outside). Each potassium ion that leaves the cell will carry positive charge with it, leaving a net negative charge behind. As more potassium ions leave the cell, charge builds up on the membrane: positive outside as the potassium ions accumulate, but negative inside because potassium ions have been lost. The separated charges cling tightly to the insulating membrane, electrically attracting each other, and set up an electrical potential (or voltage) across the membrane.

As the voltage increase builds, this potential makes it difficult for more potassium ions to travel outward, toward the positively charged side. At the same time, it becomes easier for potassium ions traveling inward, since they are heading toward negative charge. When the membrane potential reaches a certain value, the outward flow slows and the inward flow increases so much that they balance each other, and the cell stays in a steady state with a steady voltage on the membrane. This voltage is called the equilibrium potential for that particular ion. Other ions have different concentrations inside and out, so their equilibrium potentials would be different.

The equilibrium potentials for other ions depend on concentrations inside and outside the cell, and the charge of the individual ions. Sodium would therefore reach its equilibrium potential when positive voltage inside the membrane slowed its rate of inward leakage and increased its

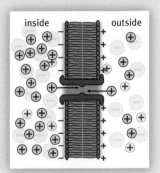

Figure A1.1
Equilibrium potential

This state would be reached if only one ion was able to cross the membrane. Ions flow from strong to weak concentration, building electric charge on the membrane.

At equilibrium, flow rates in both directions match with unequal concentrations on either side, because electrostatic forces make it harder to flow from inside to out, and easier to flow from the outside in.

outward leakage. Every kind of ion has an equilibrium potential, but whether the cell membrane actually moves toward that equilibrium depends on whether that kind of ion can pass through.

Resting membrane potential

If all the ions were free to leak across the membrane, they would flow from the concentrated side to the dilute side, so that potassium would leak out of the neuron and sodium would flow in to the neuron. However the membrane prevents them from doing so, except through specialized openings (channels), which are selective for each type of ion. At any given moment the membrane potential is influenced by which ions are able to pass through and carry charges across; as channels open and close, the membrane potential moves toward the equilibrium potential of whichever ions pass through most easily.

In a steady state, neuron membranes are permeable to potassium, which makes membrane voltage move toward the potassium equilibrium potential. However, other ions, including sodium, can also cross the membrane, even though their movement is more restricted. As the membrane approaches the potassium equilibrium potential, the tiny rate of inward sodium leakage increases and brings positive charge in, altering membrane potential the other way. The net result is that the resting membrane potential is close to the potassium equilibrium potential, but does not quite reach it. This mix between the potassium and sodium equilibrium potentials is determined by the membrane permeability for each ion. This also means that the inward and outward flows do not balance for either ion, meaning that there is a steady loss of potassium and a steady gain of sodium. The net result is that the cell has to burn energy to pump ions to maintain concentrations.

Action potential

With potassium channels open in the steady state, the membrane potential stays near the equilibrium potential for potassium, with a negative charge inside the membrane and a positive charge on the outside. Specific triggers can activate mechanisms that open sodium channels, causing the membrane potential to move rapidly toward the sodium equilibrium, for a brief period of time. This brief, clear signaling event is known as an action potential.

Because sodium ions are maintained with the opposite concentration difference–weak concentration inside, stronger concentration outside–opening sodium channels will push the membrane potential the opposite way, toward positive charge inside and negative outside. Because the amount of charge required is very small, and moves only a few nanometres, these changes can occur very rapidly. Neurons and muscle cells exploit this by creating a brief, sharp flip between two states. This spike in voltage is used to trigger signaling at the synapses between neurons. To achieve the flip, voltage-gated sodium channels need to open and close. In the signaling parts of the axon, the membrane is studded with sodium channels, which open when the membrane potential reaches a certain threshold level. Opening these channels triggers a brief, time-limited reversal of membrane potential, which stops after just a millisecond or two. Potassium channels open briefly to restore the previous state. Because the process triggers a large voltage change, it propagates itself in a spreading wave that races across the membrane, like a Mexican wave in a stadium crowd.

Many descriptions of the action potential focus on the movement of ions. They focus on ions rushing in and out of ion channels, ions being pumped by membrane mechanisms, and so on. However, the simplest way to think about action potentials is to note that the established

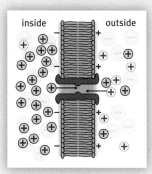

Figure A1.2
Resting potential

In the resting cell, multiple ion types continuously cross the membrane through leak channels. None of them are at equilibrium, so active pumping is required to maintain concentrations against the steady leakage. Membrane potential stabilizes in between the equilibrium potentials for each ion, according to leakage rates.

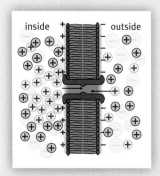

Figure A1.3
During action potential

Although the large scale composition of the fluids inside and out remains almost the same, changes in permeability allow sudden freedom for sodium ions to cross the membrane. This causes a brief reversal of the membrane potential toward the sodium equilibrium.

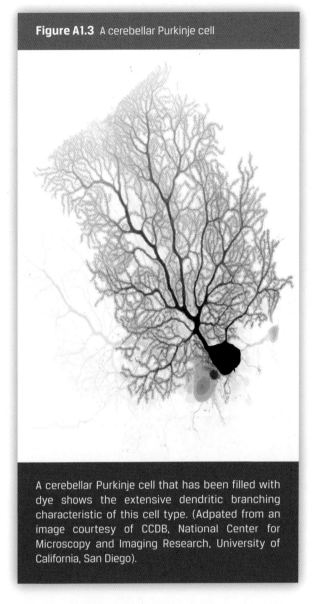

Figure A1.3 A cerebellar Purkinje cell

A cerebellar Purkinje cell that has been filled with dye shows the extensive dendritic branching characteristic of this cell type. (Adpated from an image courtesy of CCDB, National Center for Microscopy and Imaging Research, University of California, San Diego).

concentration differences of potassium and sodium will set up opposing equilibrium potentials, which are the extremes of the range of membrane potential in a normal neuron. The actual, moment-to-moment membrane potential varies within this range according to which channels are open. The threshold for triggering an action potential is a fixed voltage determined by the properties of voltage-gated channels. The events leading to an action potential can be summarized as follows:

1. At rest, the (negative) potassium equilibrium potential dominates.

2. Excitatory transmitters open sodium channels, so the membrane potential moves a little way toward the positive sodium equilibrium potential. Since this means moving from a negative membrane polarization toward zero, this is termed 'depolarization.'

3. If this takes the membrane potential past threshold, many more sodium channels open and the membrane potential jumps much closer to the sodium equilibrium potential; this creates a voltage spike.

4. After this, the sodium channels are abruptly closed, and potassium channels are opened, so membrane potential jumps back to the negative potassium equilibrium.

5. All the channels are reset and the cell returns to the resting, potassium-dominated state.

Many ions other than potassium and sodium also move in or out of the cell during these processes, but the main point to grasp is that the sodium-potassium pump works to keep the equilibrium potentials stable and the cell operating consistently. Note that inhibitory transmitters hyperpolarize the membrane (they increase the negative charge inside the cell) and make the production of an action potential less likely.

Appendix B

Changing names–worn out terms and confusing categories

This appendix brings together the story of a number of terms that are commonly used, but which we believe are a cause of confusion. These terms appear in most textbooks and are in common use by neurologists, but it is time that their impact was critically examined. Each of these topics is dealt with in main body of the text, but they are presented here for the convenience of the reader.

As knowledge of the brain developed over the last century, scientists and clinicians searched for terms that would simplify the growing complexity and make it understandable to a wider audience. Some of these new terms became popular even though they were not supported by evidence. Brodal pointed out that this very human tendency to replace mystery with terminology was well described by Goethe: "Just at the point where human comprehension fails, a word steps in to neatly fill its place."

Now that we know much more about the brain, we think it is time to abandon some of these terms, and to use others only in a restricted sense. The following are the terms we consider to be problematical.

- The limbic system
- The reticular activating system
- The extrapyramidal system
- The rhinencephalon
- The basal ganglia
- The pons as a subdivision of the hindbrain

The limbic system

The term 'limbic system' has its origin in 'le grand lobe limbique' of Broca, who drew attention to the curved belt of telencephalic structures that border the central parts of the forebrain. This definition of the 'limbic lobe' includes the hippocampus, the amygdala, the cingulate gyrus, and the fornix. However, in the contemporary literature, the limbic lobe is often said to include the various parts of the hippocampal formation, the septum, the olfactory tubercle, the bed nucleus of the stria terminalis, the amygdala, and the cingulate gyrus. Different authors add or subtract structures from the definition in an seemingly random manner. One reason for the popularity of the term 'limbic system' is that scientists have wanted to find a term within which to group a heterogeneous collection of basal forebrain structures that are difficult to classify. But as Brodal critically observed, "It is difficult to see that the lumping together of these different regions under one anatomical heading, 'the limbic lobe' serves any purpose".

Brodal argues that the use of the term 'limbic system' is even more dangerous, because the supposed components do not share a common function. Instead, each has a distinct set of functions, ranging from olfaction to emotional responses to memory registration. In an attempt to deal with the diversity of structures thrown in to the 'limbic system' Mesulam and colleagues have (1977, p 407) suggested that the definition of 'limbic' should include all centers that are either directly or indirectly connected with the hypothalamus. This is similar to the attempts to make the term limbic system equivalent to the now discredited term 'visceral brain'. If this approach is taken to extremes, then almost everything in the central nervous system can be called limbic.

The answer to these increasingly confusing attempts to defend an unfortunate name is to abandon the term 'limbic' altogether. Instead, one should speak specifically about hippocampus, amygdala, septum, or whatever part of the brain is under discussion, without connection to the baggage created by the term 'limbic'.

The reticular activating system

The centers to keep the brain awake were originally described by physiologists as the reticular activating system of the brainstem. Their experiments showed that ascending pathways from the brainstem caused the cerebrum to be alert, and the brainstem nuclei that gave rise to the ascending pathways were thought to reside in the so-called reticular core of the brainstem. Unfortunately, there was confusion over the identity of the brainstem centers that promoted wakefulness, and only recently have the specific nuclei been identified. Since the nuclei that promote wakefulness (the locus coeruleus, the raphe nuclei, and the major cholinergic nuclei of the hindbrain) are not part of the true reticular formation of the brainstem, the term 'reticular activating system' is no longer useful and should be replaced by 'ascending arousal system'.

The extrapyramidal system

This term became popular among neurologists who were trying to distinguish the motor problems arising from damage to the pyramidal (corticospinal) tract from other motor syndromes. They contrasted the direct (monosynaptic) projection of the cortex to motoneurons (the pyramidal system) with what they believed to be an alternative multisynaptic pathway from cortex to motoneurons. They hypothesized that this multisynaptic pathway went from cortex to striatum, pallidum, various midbrain and hindbrain centers, and finally to spinal cord. The problem with this concept is that the hypothesized chain of connections does not exist. In 1962, Nauta and Mehler showed that the globus pallidus projects almost entirely to the thalamus rather than to midbrain and hindbrain motor centers. Despite this, the 'extrapyramidal' concept lives on in textbooks and in lectures given by neurologists.

Rhinencephalon

This word means 'nose brain'. It was used to describe the phylogenitically older parts of the forebrain that are related to olfaction. Unfortunately, the use of the term was expanded to include other phylogenetically old parts of the forebrain that have nothing in particular to do with olfaction, such as the hippocampus. Because of the confusion created by its use, the term 'rhinencephalon' is one to avoid.

Basal ganglia

This term is most commonly used in clinical settings to refer to the combination of the striatum and pallidum, because both are subpallial structures and they have related motor functions. This usage is basically acceptable, but many people use the term in different ways. For example, some add the amygdala to the striatum and pallidum to make up the basal ganglia, because the the amygdala shares its subpallial origins with the striatum and pallidum, even though the amygdala is functionally quite distinct. Others take a different approach and use the term to include the subthalamic nucleus and substantia nigra with the striatum and pallidum, on the basis that they all have related motor functions. The problem with this approach is that the subthalamus and substantia nigra are not subpallial structures, and have no logical place in a group with the striatum and pallidum. To make matters worse, much older definitions use the term to include the thalamus with the striatum and pallidum. For all these reasons, it is a term to use with care. In most cases it is better to refer directly to the striatum or pallidum or both, and avoid the use of the term 'basal ganglia'.

The pons as a subdivision of the hindbrain

It is customary for human anatomy textbooks to divide the hindbrain into a rostral part, the pons, and a caudal part, the medulla oblongata. Recent gene expression data on the development of the hindbrain shows us that this usage will have to be abandoned. The pons certainly exists; it is a large group of nuclei in the rostral hindbrain. The pontine nuclei connect the cerebral cortex with the cerebellum. In humans, the pontine nuclei are huge, and cover about half of the ventral surface of the hindbrain, down to the caudal end of the facial nucleus. However, in the rat and mouse, the pontine nuclei are of modest size and do not extend to cover the facial nucleus. The fundamental reason for this difference is found the early development of the hindbrain. The pontine nuclei arise from the dorsal surface of the sixth rhombomere (r6) and migrate to the ventral surface of rhombomeres three and four (r3 and r4), where they start to expand. Even in humans, where the expansion is massive, the pontine nuclei remain anchored to r3 and r4, and are never joined to the rostral hindbrain (isthmus, r1, r2) or the hindbrain near the facial nucleus (r5, r6). Because the extent of the pons varies from species to species, it is not a reliable or logical way to define the rostral hindbrain. The term 'pons' should be used only in reference to the pontine nuclei and their related fiber bundles.

The medulla oblongata as a subdivision of the hindbrain

Because the definition of the pons as a subdivision of the hindbrain is problematic, the definition of medulla oblongata is also compromised. The human medulla oblongata (medulla) is defined by the fact that it is not the pons. Since the relative size of the pons changes from species to species, the best way to avoid further confusion is to call the medulla the caudal hindbrain. There is no doubt that the term 'medulla' will still be used as a synonym for the caudal hindbrain, but students should be aware of the limitations of this usage.

Appendix C

Notes on the images beginning each chapter.

Chapter 1

This is a photomicrograph of a rat spinal cord at level L5 stained with NeuN. In this picture the colors have been inverted for dramatic effect, so that the normally dark staining neurons appear white on a dark background.

Chapter 2

This is a diffusion tensor image (DTI) of a mouse brain. In this image the mathematical 'seed' has been placed in the motor cortex. The commissural connections with the opposite hemisphere can be seen in green and the descending pathways are represented in blue. Connections with association motor areas are shown in red. This image was provided by Randal Moldrich and Nyoman Kurniawan, Queensland Brain Institute, Brisbane, Australia.

Chapter 3

This is a photograph of a sagittal section of rat brain, stained with the Nissl method.

Chapter 4

A photograph of a dissection of human spinal cord. The dorsal surface of the spinal cord is visible and the dorsal and ventral rootlets can be seen as they form the dorsal and ventral roots of each spinal nerve. This image was provided by of Professor Mete Erturk, Department of Anatomy, Ege University, Turkey.

Chapter 5

This is a photograph of fighting kangaroos at the Tama Zoo in Tokyo. Male kangaroos fight for dominance by scraping their powerful hind legs against the abdomen of their opponent. They balance on their tail while doing this. The photograph was taken by Dr Hironobu Tokuno, of the Tokyo Metropolitan Institute for Neuroscience, Japan.

Chapter 6

This is a photomicrograph of a thin section of a monkey retina. The different layers of retinal cells are clearly visible. Photoreceptors are located adjacent to the choroid layer (bottom of the picture), which synapse with bipolar cells that transmit information to the optic nerve via the ganglion cells on the surface of the retina (top of the picture). This image was provided by Professor Jan Provis, Australian National University, Canberra, Australia.

Chapter 7

This is a magnetic resonance image of the head. Magnetic resonance produces images based on water molecules in the tissue. The face looks a little strange because the front of the nose is not included in this image. The main areas of the forebrain and the surrounding skull bones can be seen in this MR representation of a horizontal slice.

Chapter 8

This is a diffusion-rendered image of a magnetic resonance imaging datasheet, provided by Dr David S. Ebert, Director of the Purdue Visualization and Analytics Center at Purdue University.

Chapter 9

A fluorescent labeled section of a human brain, showing some of the hallmarks of Alzheimer's disease. The plaques of insoluble amyloid beta protein are stained blue by thioflavine S. Axons in the surrounding neuropil contain neurofilament proteins (green), which are disrupted into blobs, rings and clumps near the amyloid plaques. Inhibitory neurons containing the protein calretinin (red) appear unaffected. Lipofuscin, a fatty deposit in neurons, appears as white blobs. (Image courtesy Stanislaw Mitew, University of Tasmania, Australia).

Chapter 10

This is an image of a developing mouse brain showing one of the migrations from the ganglionic eminence at embryonic day 16 (E16). Migrating calretinin-positive interneurons migrating on the surface of the cerebrum appear greenish-yellow in this fluorescence image. This superficial migration is secondary to the main tangential migration of interneurons into the cerebrum which occurs earlier than E16. The dense oval concentration of calretinin fluorescence on the lower right hand side is in the position of the lateral olfactory tract. The blue stained areas are stained with DAPI, which shows the presence of actively dividing cells. This image was provided by Dr Stacey Cole of the Queensland Brain Institute, Brisbane, Australia.

Chapter 11

A rat brain section which has undergone a modified Gallyas stain, which deposits silver on myelin sheaths. The thick section (100 micrometers) allows cortical structure to be visualized.

Supplementary reading

General books on neuroanatomy

Bear MF, Connors BW, Paradiso MA (2007) *Neuroscience: exploring the brain.* 3rd Edition. Lippincott, Williams and Wilkins: New York.

Brodal P (2010) *The Central Nervous System: structure and function.* 4th Edition. Oxford University Press: New York.

Swanson LW (2003) *Brain Architecture: understanding the basic plan.* Oxford University Press: New York.

Chapter One–Nerve cells and their synapses

Membrane potentials and action potentials

Kandel ER, Schwartz JH, Jessell TM (2000) *Principles of Neural Science.* 4th Edition. McGraw-Hill: New York.

Neurons and their connections

Bear MF, Connors BW, Paradiso MA (2007) *Neuroscience: exploring the brain.* 3rd Edition. Lippincott, Williams and Wilkins: New York. *(Probably the most widely used introductory text in neuroscience. Authoritative and well written.)*

Glia

Kettenman H, Ransom B (1995) *Neuroglia.* Oxford University Press: Oxford.

Chapter Two–Central nervous system basics–the brain and spinal cord

External features of the brain

Brodal P (2010) *The Central Nervous System: structure and function.* 4th edition. Oxford University Press: New York. *(A modern version of the classic text written by the author's father, Alf Brodal)*

Bowsher D (1967) *Introduction to the Anatomy and Physiology of the Nervous System.* Blackwell Scientific Publications: Oxford. *(A little gem–concisely written but very understandable).*

The spinal cord

Watson C, Paxinos G, Kayalioglu G (eds) (2009) *The Spinal Cord. A Christopher and Dana Reeve Foundation Text and Atlas.* Elsevier Academic Press: San Diego.

Meninges, ventricles, and circulation of cerebrospinal fluid

Brodal P (2004) *The Central Nervous System: structure and function.* 3rd edition. Oxford University Press: New York.

Chapter Three–A map of the brain

Mini-atlas of the rat brain

Paxinos G, Watson C (2007) *The Rat Brain in Stereotaxic Coordinates.* 6th Edition. Elsevier

Academic Press: San Diego. *(The most cited work in neuroanatomy–an essential reference for researchers)*.

Franklin KBJ, Paxinos G (2007) *The Mouse Brain in Stereotaxic Coordinates.* 3rd Edition. Academic Press: San Diego.

Mai JK, Paxinos G, Voss T (2008) *Atlas of the Human Brain.* 3rd Edition. Academic Press: San Diego.

Chapter Four–The peripheral nerves

Spinal nerves supplying the limbs

Cramer GD, Darby SA (2005) *Basic and Clinical Anatomy of the Spine, Spinal cord and ANS.* 2nd Edition. Elsevier Mosby: Missouri.

Watson C, Paxinos G, Kayalioglu G (eds) (2009) *The Spinal Cord. A Christopher and Dana Reeve Foundation Text and Atlas.* Elsevier Academic Press: San Diego.

Autonomic nervous system

Cramer GD, Darby SA (2005) *Basic and Clinical Anatomy of the Spine, Spinal cord and ANS.* 2nd Edition. Elsevier Mosby: Missouri.

Furness JB (2006) *The Enteric Nervous System.* Wiley-Blackwell: Oxford.

Chapter Five–Command and control–the motor systems

Command and control of skeletal muscles

Rizzolatti G, Craighero L (2004) The mirror-neuron system. Annu Rev Neurosci 27, 169-92

Desmurget M, Reilly KT, Richard N, Szathmari A, Mottolese C, Sirigu A (2009) Movement intention after parietal cortex stimulation in humans. Science 324, 811-3 *(Direct evidence of the role of parietal cortex in forming targets for movement)*

The corticospinal tract

Lemon RN, Johansson RS, Westling G (1995) Corticospinal control during reach, grasp, and precision lift in man. J Neurosc 15, 6145-6156

Survival skills: the hypothalamus

Swanson LW (2003) *Brain Architecture: understanding the basic plan.* Oxford University Press: New York.

Brainstem and spinal cord modules for control of organized movement

Watson C, Paxinos G, Kayalioglu G (eds) (2009) *The Spinal Cord. A Christopher and Dana Reeve Foundation Text and Atlas.* Elsevier Academic Press: San Diego. *(A comprehensive reference for the anatomy of the descending tracts and motor neuron groups)*

The role of the cerebellum in motor control

Glickstein R (2007)What does the cerebellum really do? Current Biol 17: R824-R827

Paxinos G (2004) *The Rat Nervous System.* 3rd Edition. Elsevier Academic Press: San Diego.

Shambes GM, Gibson JM, Welker W (1978) Fractured somatotopy in granule cell tactile areas of rat cerebellar hemispheres revealed in micromapping. Brain Behav Evol 15, 94-140.

The roles of the striatum and pallidum in motor control

Paxinos G, Watson C (2007) *The Rat Brain in Stereotaxic Coordinates.* 6th Edition. Elsevier Academic Press: San Diego.

Command and control of the visceral systems–the autonomic nervous system

Furness JB (2006) *The Enteric Nervous System*. Wiley-Blackwell: Oxford.

Chapter Six–Gathering vital information–the sensory systems

Keeping sensory maps intact

Woolsey TA, Van der Loos H (1970) The structural organization of layer IV in the somatosensory region (S1) of the mouse cerebral cortex. Brain Research 17, 205-242.

Somatosensory system

Paxinos G, Watson C (2007) *The Rat Brain in Stereotaxic Coordinates*. 6th Edition. Elsevier Academic Press: San Diego.

Visual system

Mai JK, Paxinos G (2004) *The Human Nervous System*. 2nd Edition. Elsevier Academic Press: San Diego.

Paxinos G (2004) *The Rat Nervous System* 3rd Edition. Elsevier Academic Press: San Diego.

Zeki S (1993) *A Vision of the Brain*. Blackwell Scientific: Oxford.

Olfaction

Nieuwenhuys R, Ten Donkelaar HJ, Nicholson C (1998) *The central nervous system of vertebrates*. Vol 3. Springer, New York.

Chapter Seven–The human cerebral cortex

The cerebral cortex–anatomy and histology

Herculano-Houzel S (2009) The human brain in numbers: a linearly scaled-up primate brain. Human Neurosc 3, 1-11.

Guiding principles of cortical structure and function

Ramachandran VS, Blakeslee S (1999) *Phantoms in the Brain*. Harper: New York

The functional layout of human cerebral cortex

Disbrow E, Hinckley L, Krubitzer L, Padberrg J (2007) Handuse and the evolution of posterior parietal cortex in primates. In *Evolution of Nervous Systems*. Volume 4–Primates. Kaas JH and Preuss TM (eds) pp407-415. Elsevier Academic Press: San Diego.

Dum RP, Strick PL (2002) Motor areas in the frontal lobe of the primate. Physiol Behav 77, 677-82.

Graziano MSA, Taylor CS, Moore T (2002) Complex movements evoked by microstimulation of precentral cortex. Neuron 34, 841-851.

Kaas JH (2007) The evolution of sensory and motor systems in primates. In *Evolution of Nervous Systems*. Volume 4–Primates. Kaas JH and Preuss TM (eds) pp35-57. Elsevier/Academic Press: San Diego.

Rizzolatti G, Arbib MA (1998) Language within our grasp. Trends Neurosci 21, 188-194.

Mai JK, Paxinos G, Voss T (2008) *Atlas of the Human Brain*. 3rd Edition. Academic Press: San Diego.

Mai JK, Paxinos G (2004) *The Human Nervous System*. 2nd Edition. Elsevier Academic Press: San Diego.

Wise SP (2007) The evolution of sensory and motor systems in primates. In *Evolution of Nervous Systems*. Volume 4–Primates. Kaas JH and Preuss TM Eds. pp157-165. Elsevier/Academic Press: San Diego.

The cerebral cortex and behavior

Barrett L, Henzi P (2005) The social nature of primate cognition. Proc. R. Soc. B 272, 1865-1875.

Kirkcaldie MTK, Kitchener PD (2007) When brains expand: mind and the evolution of the cortex. Acta Neuropsych 19, 139-148.

Chapter Eight–High level functions-consciousness, memory, learning, and emotions

Consciousness

Sergent C, Baillet S, Dehaene S (2005) Timing of the brain events underlying access to consciousness during the attentional blink. Nature Neurosc 8, 1391-1400.

Dimond S (1972) *The Double Brain*. The Whitefriars Press Ltd, London.

Gazzaniga MS, Sperry RW (1966) Simultaneous double discrimination response following brain bisection. Psychon Sci 4, 261-262.

Saper CB, Chou TC, Scammell TE (2001) The sleep switch: hypothalamic control of sleep and wakefulness. Trends Neurosci 24, 726-731. *(The story of the discovery of the sleep switch, written by the team who made the discovery).*

Saper CB, Scammell TE, Lu J (2005) Hypothalamic regulation of sleep and circadian rhythms. Nature 437: 1257-63.

Diekelmann S, Born J (2010) The memory function of sleep. Nat Rev Neurosci 11, 114-26.

Siegel JM (2009) Sleep viewed as a state of adaptive inactivity. Nat Rev Neurosci 10, 747-53.

Memory

Anderson P, Morris R, Amaral D, Bliss T, O'Keefe J (eds) (2007) *The Hippocampus Book*. Oxford University Press: New York.

Dan Y, Poo MM (2006) Spike timing-dependent plasticity: from synapse to perception. Physiol Rev 86, 1033-48

Citri A, Malenka RC (2008) Synaptic plasticity: multiple forms, functions, and mechanisms. Neuropsychopharm 33, 18-41

Tailby C, Wright LL, Metha AB, Calford MB (2005) Activity-dependent maintenance and growth of dendrites in adult cortex. Proc. Natl. Acad. Sci. U.S.A. 102 4631-6.

Rietze RL, Valcanis H, Brooker GF, Thomas T, Voss AK, Bartlett PF (2001) Purification of a pluripotent neural stem cell from the adult mouse brain. Nature 412, 736-739.

Bull ND, Bartlett PF (2005) The adult mouse hippocampal progenitor is neurogenic but not a stem cell. J Neurosci 25, 10815-10821.

Emotions and the amygdala

LeDoux J (1998) *The emotional brain: the mysterious underpinnings of emotional life*. Simon and Schuster: New York. *(LeDoux is a fine writer with excellent research credentials. This is a very good read).*

LeDoux J (2007) The amygdala. Current Biology 17, R868-874. *(An excellent review of a complex subject).*

Swanson LW, Petrovich GD (1998) What is the amygdala? Trends Neurosci. 21, 323-331. *(This is the definitive review of the anatomy of the amygdala and its connections.)*

Limbic system

Brodal A (1981) *Neurological Anatomy in relation to clinical medicine.* 3rd Edition. Oxford University Press: London. *(This is an excellent textbook by an author who understood both laboratory neuroanatomy and clinical neurology).*

Kötter R, Stephan KE (1997) Useless or helpful? The concept "limbic system". Rev in Neurosci 8, 139-145.

Chapter Nine–When things go wrong–brain disease and injury

Donaghy M (2009) *Brain's Diseases of the Nervous System.* 12th Edition. Oxford University Press: New York. *(This classic textbook, originally written by the late Lord Russel Brain (could he have been anything else but a neurologist!) is the standard reference for neurological illness).*

Rhodes R (1997) *Deadly feasts.* Simon and Schuster: New York. *(A brilliant account of prion disease, Kuru, and mad cow disease that reads like a detective story).*

The tragic history of treatment of severe mental illness

Scull A (2006) *The Insanity of Place/The Place of Insanity: Essays on the History of Psychiatry.* Routledge: London.

Scull A (2007) *Madhouse: A Tragic Tale of Megalomania and Modern Medicine.* Yale University Press: New Haven.

Valenstein ES (1973) *Brain Control: A Critical Examination of Brain Stimulation and Psychosurgery.* John Wiley and Sons Inc: New York. *(This book had a powerful influence on attitudes to psychosurgery and helped to stop the excessive use of these techniques).*

Valenstein ES (1986) *Great and Desperate Cures: The Rise and Decline of Psychosurgery and other Radical Treatments for Mental Illness.* Basic Books: New York.

Chapter Ten–The development of the brain and spinal cord

Genes and brain development

Rallu M, Corbin JG, Fishell G (2002) Parsing the telencephalon. Nat Rev Neurosci 3, 943-950 *(An outstanding account of the role of genes in forebrain development)*

Allman JM (1998) *Evolving Brains.* Scientific American Library: New York. *(A wonderfully readable account of the evolution of mammalian brains.)*

Early development of the brain and spinal cord

Sanes D, Reh TA, Harris WA (2005) *Development of the Nervous System.* 2nd Edition. Elsevier Academic Press: San Diego. *(A textbook that should be on the shelf of everyone working in the area of brain development).*

Ashwell KWS, Paxinos G (2008) *Atlas of the Developing Rat Nervous System.* 3rd Edition. Academic Press: San Diego. *(The most accurate and comprehensive atlas of prenatal brain development)*

Later processes that refine the structure of the brain

Doidge N (2008) *The brain that changes itself*. Scribe Publications: Carlton North.

Draganski B, Gaser C, Kempermann G, Kuhn HG, Winkler J, Büchel C, May A (2006) Temporal and spatial dynamics of brain structure changes during extensive learning. J Neurosci 26, 6314-6317.

Paxinos G, Halliday G, Watson C, Koutcherov Y, Wang HQ (2007) *Atlas of the Developing Mouse Brain at E17.5, P0, and P6*. Academic Press: San Diego. *(This is particularly useful for molecular biologists studying gene expression in late term or newborn mice).*

Chapter Eleven–Techniques for studying the brain

Cutting thin sections of the brain
Woods AE, Ellis RC (1994, 1996) *Laboratory histopathology: a complete reference*. Vols. 1, 2. Churchill Livingstone: Edinburgh.

Staining brain sections
Carter M, Shieh J (2009) *Guide to Research Techniques in Neuroscience*. Elsevier Academic Press: San Diego.

Paxinos G, Watson CRR, Carrive P, Kirkcaldie MTK, Ashwell KWS (2009) *Chemoarchitectonic Atlas of the Rat Brain*. 2nd edition. Elsevier Academic Press: San Diego.

Watson CRR, Paxinos G (2010) *Chemoarchitectonic Atlas of the Mouse Brain*. Elsevier Academic Press: San Diego.

Cell culture
Shahar A, de Vellis J, Vernadakis A, Haber B (eds.) (1989) *A Dissection and Tissue Culture Manual of the Nervous System*. Alan R Liss Inc: New York.

State-of-the-art protocols from the Cold Spring Harbor research group can be accessed online at http://cshprotocols.cshlp.org/

Hodology: using tracers to map connections
Zaborszky L, Wouterlood FG, Lanciego JL (2006) *Neuroanatomical Tract-Tracing: Molecules, Neurons, and Systems*. Springer: New York.

Oztas E (2003) Neuronal tracing. Neuroanatomy 2, 2-5. *(available online at neuroanatomy.org)*

Paxinos G, Watson C (2007) *The Rat Brain in Stereotaxic Coordinates*. 6th Edition. Elsevier Academic Press: San Diego.

Molecular genetics
Pourquie O (ed) (2009) *Hox Genes*. Elsevier Academic Press: San Diego.

Non-invasive imaging techniques
Turner R, Jones T (2003). Techniques for imaging neuroscience. Br Med Bull 65, 3-20.

Electrophysiology
MDS Analytical Technologies (2008) *The Axon Guide: A Guide to Electrophysiology & Biophysics Laboratory Techniques* (available online at www.moleculardevices.com/pages/instruments/axon_guide.html)

Glossary

A

Abducens nerve: the sixth cranial nerve, which supplies the lateral rectus muscle of the eye.

Accessory nerve: the eleventh cranial nerve, which supplies the large muscles of the neck (spinal part) and contributes to the vagus nerve (cranial part).

Accumbens nucleus: a ventral part of the striatum complex, lying in the basal forebrain close to the midline. It receives an important dopamine pathway from the ventral tegmental area.

Action potential: a brief electrical signal in a neuron, created by the movement of ions across the cell membrane.

Alzheimer's disease: a degenerative brain disease that commonly affects the elderly. Patients typically have problems with memory and language, and are disoriented in time and place.

Amygdala: an almond-shaped nucleus located in the anterior temporal lobe, thought to be involved in emotional responses and social behaviors.

Anterior: towards the front of the body.

Astrocyte: a type of glial cell in the central nervous system. It is involved in controlling the cell environment, including nutrients, water balance and pH

Auditory system: the brain structures involved in the process of hearing.

Autism: a neurodevelopmental disorder that becomes apparent in childhood. Autistic children demonstrate difficulties in emotional interaction and communication.

Autoimmune: a type of illness in which the immune system attacks otherwise normal cells in the body.

Autonomic nervous system: the part of the nervous system that controls the activity of the visceral organs, blood vessels, and glands.

Axon: a elongated branch of a neuron which conducts action potentials away from the nerve cell body.

B

Basal ganglia: a term describing a group of nuclei including the caudate nucleus, putamen, and globus pallidus. These nuclei have a major role in motor control. Note that some people use this term to include the subthalamus and substantia nigra; we recommend against this. Some authors also include the amygdala in this category.

Bed nucleus of the stria terminalis: a group of nuclei near the anterior commissure that receive fibers from the amygdala. These fibers travel along a tract called the stria terminalis until they reach the bed nucleus.

Bipolar affective disorder: a psychiatric condition characterized by alternating periods of depression and mania.

Bötzinger complex: the Bötzinger nucleus and the pre-Bötzinger nucleus together represent the center for generation of respiratory rhythm in the brain. Located in the caudal hindbrain close to the ambiguus nucleus, this nuclear complex is essential for survival. When it is damaged, the breathing becomes irregular, and finally stops.

Brainstem: a term used to describe the midbrain and hindbrain together. Because there is no consensus on the proper use of this term, it is better to refer to the midbrain and hindbrain individually wherever possible.

Branchial arch: the branchial arches are solid structures that separate the gill slits of fish. The remnants of the branchial arches are still present in mammalian embryos, forming parts of the head and neck. In mammals the first branchial arch forms the jaw bone and chewing muscles.

Branchial motor: the muscles of the branchial arches are supplied by branchial motor nerves. The branchial motor nerve fibers are present in the trigeminal, facial, glossopharyngeal, vagus, and accessory nerves.

C

Caudal: towards the tail.

Central nervous system: forebrain, midbrain, hindbrain (including the cerebellum), and spinal cord.

Cerebellum: a dorsal expansion of the hindbrain, essential for motor coordination and posture.

Cerebral cortex: the layer of gray matter on the surface of the cerebrum.

Cerebral hemisphere: the cerebrum is divided into two hemispheres, the left and the right, by the longitudinal cerebral fissure.

Cerebrospinal fluid: a clear watery fluid that surrounds and fills the ventricles of the brain.

Cerebrum: the large domed area of the forebrain consisting of an outer layer of cortex and a core of white matter. The cerebrum consists of two cerebral hemispheres, one on each side.

Cingulum: a large bundle of fibers that runs longitudinally along the dorsal surface of the corpus callosum, deep to the cingulate gyrus.

CNS: see central nervous system.

Commissure: a bundle of axons that crosses the midline to connect similar regions on different sides of the brain.

Corpus callosum: the major cerebral commissure. The fibers of the corpus callosum connect the two hemispheres of the cerebral cortex.

Cortex: see cerebral cortex.

Corticospinal tract: a bundle of axons which travels from the motor cortex to the spinal cord. It is important in the control of voluntary movement in humans.

Cranial nerves: the twelve pairs of nerves arising from the brain, responsible for sensation and motor control of the head and neck, and some viscera.

Creutzfeldt-Jakob Disease: a prion disease that causes rapid and fatal brain degeneration. Patients may experience memory problems, bizarre behavior, and motor abnormalities.

Critical period: a period in early development where environmental stimuli can irreversibly alter future development.

CSF: see cerebrospinal fluid.

D

Decussation: axons that cross the midline while traveling between regions of the CNS.

Demyelination: loss of myelin sheaths that occurs in certain neurological conditions, such as multiple sclerosis.

Dendrite: projections of the nerve cell body that look like the branches of a tree. Dendrites receive input connections from axon terminals of other neurons.

Diencephalon: one of the major subdivisions of the forebrain. The parts of the diencephalon are the pretectum (prosomere 1), the thalamus (prosomere 2), and the prethalamus (prosomere 3). The hypothalamus was formerly included in the diencephalon but gene expression analysis shows that it is a distinct entity.

Dorsal: towards the back.

Dorsal column-medial lemniscus pathway: the main touch pathway from spinal cord to thalamus.

Dorsal horn: the dorsal area of gray matter of the spinal cord that receives information from the sensory roots of spinal nerves.

Dorsal root ganglion: a swelling on the dorsal root of a spinal nerve that contains the cell bodies of the primary sensory neurons.

E

Effector system: any system which produces actions such as muscle contraction, secretion of glands, or release of hormones.

Embryonic period: the period of early development when the majority of the body's vital structures are formed, corresponding to the first eight weeks of pregnancy in humans.

Endocrine system: the system that secretes hormones into the blood stream. Hormones have effects on distant organs or cells.

Enteric nervous system: a division of the autonomic nervous system consisting of the network of neurons in the wall of the digestive tract.

Entorhinal cortex: an area of cortex in the temporal lobe close to the hippocampus that encodes a spatial map of the external surroundings and sends this information to the hippocampus.

F

Facial nerve: the seventh cranial nerve. It supplies the muscles of the face and the sensation of taste to the front of the tongue.

Fasciculus: a bundle of nerve axons.

Fetus: the name for the developing baby after it has passed the embryonic stage of development. The alternative spelling 'foetus' should be avoided in scientific writing.

Forebrain: the most rostral part of the brain, also called the prosencephalon. The parts of the forebrain are the cerebral cortex (called the pallium in the embryo), the subpallium, the hypothalamus, the eye, and the diencephalon.

G

Ganglion: a collection of neuron cell bodies in the peripheral nervous system.

Gene: a long sequence of DNA that carries the instructions for making a particular polypeptide or protein, or regulates the activity of other genes.

Glia: see Neuroglia

Globus pallidus: part of the pallidum. It receives input from the striatum and projects mainly to the thalamus.

Glossopharygneal nerve: the ninth cranial nerve, which supplies the muscles of the pharynx, the salivary glands and, sensation to the back part of the tongue.

Gray matter: any part of the central nervous system containing many neuronal cell bodies, dendrites, and glia.

Gyrus: a bulge on the surface of the cerebral cortex. The depression between two gyri is called a sulcus or a fissure.

H

Hindbrain: the part of the brainstem caudal to the midbrain. It consists of the isthmus and eleven rhombomeres. The cerebellum is a dorsal expansion of the hindbrain.

Hippocampus: an area of the temporal lobe of the brain involved in memory.

Hypoglossal nerve: the twelfth cranial nerve, which supplies the muscles of the tongue.

Hypothalamus: a major subdivision of the forebrain located ventral to the thalamus. It is responsible for the initiation of survival behaviors and controls the autonomic and neuroendocrine systems.

I

Immunohistochemistry: the use of antigen/antibody reactions to mark specific neuronal cell bodies or processes, linked to a label such as diaminobenzidine (DAB) to produce a visible product.

Inferior colliculus: a raised area on the dorsal surface of the midbrain, involved in processing auditory signals.

Inferior olive: a nucleus in the ventral part of the hindbrain whose axons project to the cerebellar cortex.

Insula: the area of cerebral cortex that receives taste sensation. In humans this area is an island of cortex found in the depths of the lateral fissure.

Isthmus: the most rostral part of the hindbrain, joining the midbrain to the rhombomeres.

L

Lateral: away from the midline.

Lemniscus: a large bundle of axons.

M

Mammillary body: a group of nuclei in the caudal part of the hypothalamus. They are involved in exploratory and foraging behaviors.

Medial: closer to the midline.

Medulla: an abbreviation of the term medulla oblongata, the caudal part of the hindbrain. In humans, the medulla is defined as the part of the hindbrain that is caudal to the pons.

Membrane potential: a voltage difference between the inside and outside of a cell.

Memory: retention of learned information or experience.

Mesencephalon: the region of the central nervous system between the forebrain and the hindbrain; also known as the midbrain.

Motor cortex: the cortical area most intimately related to execution of movement. It is located in the pre-central gyrus in humans.

Motor nerve: a bundle of axons whose terminals synapse with muscle cells and can therefore cause muscle contraction.

Multiple sclerosis: a degenerative disease of the nervous system that is characterized by patches of inflammation and demyelination of axons. Symptoms include blindness, muscle weakness, and abnormal sensation.

N

Neocortex: the part of the cerebral cortex that is greatly developed in mammals. It contains six layers and can also be called the isocortex.

Neural crest: a column of cells which splits off from the developing neural tube, forming the ganglia of the peripheral nervous system, the melanocytes of the skin, and the bones and muscles of the face.

Neural tube: a cylinder of cells that develops to form the central nervous system.

Neural tube defects: any congenital abnormality caused by defective closure of the neural tube during fetal development.

Neuroendocrine system: a system in which neurotransmitters are released as hormones into the blood stream.

Neuroglia: regulatory cells in the nervous system. The main types are astrocytes, oligodendrocytes, microglia, and Schwann cells.

Neuropil: the term neuropil describes the space between cell bodies as seen in histological sections. The neuropil is full of axons, dendrites, synapses, and glial cells, but these structures are not seen in standard histological preparations such as Nissl stained sections.

Neurotransmitter: a chemical released from nerve endings into the synaptic cleft. The neurotransmitter binds to receptors on nerve membranes to affect their function.

Neuron: a nerve cell. The neuron is the basic functional unit of the nervous system.

Nissl stain: a stain using basic dyes that bind to RNA and some DNA in cell bodies.

Nociception: the detection of potentially harmful stimuli.

Nucleus: this word has two meanings: the DNA-containing part of a cell; or in the central nervous system, a collection of neuron cell bodies.

O

Oculomotor nerve: the third cranial nerve, which supplies the muscles that move the eye and the eyelid, and controls the pupil and the lens.

Olfaction: the sense of smell.

Olfactory bulb: a cylindrical region lying on top of the nasal cavity. It is the main receiving center for the sense of smell. It is found at the rostral end of the brain in rodents, but on the ventral surface of the brain in humans.

Olfactory nerve: the first cranial nerve, which is involved in the sensation of smell.

Oligodendrocyte: a type of glial cell in the central nervous system which forms myelin sheaths to insulate axons in the brain.

Optic chiasm: the crossing of optic nerve fibers. It lies ventral to the rostral part of the hypothalamus.

Optic nerve: the second cranial nerve, which is responsible for vision.

Optic tract: a bundle of fibers carrying visual information from the optic chiasm to the lateral geniculate nucleus of the thalamus and the superior colliculus.

Organ of Corti: the auditory receptor apparatus of the inner ear.

P

Pain: a distressing mental state often triggered by nociception

Pallidum: a mass of gray matter within the cerebrum involved in motor control. The main part of the pallidum is called the globus pallidus.

Pallium: the embryonic precursor of the cerebral cortex. The word pallium means a cloak or a covering.

Palsy: palsy is an old word for paralysis. It is still commonly used by neurologists, who use facial palsy as the name for facial nerve paralysis. Parkinson's disease was originally called the 'shaking palsy', because of the combination of tremor and movement restriction.

Paracrine: a method of signalling in the body where neurotransmitter chemicals secreted by cells directly affect other nearby cells.

Parasympathetic nervous system: a division of the autonomic nervous system that controls the normal function of the internal organs and glands.

Parkinson's Disease: a brain disease caused by cell death in the substantia nigra leading to a loss of dopamine in the nigrostriatal pathway. Patients experience many motor symptoms including slow movements, tremor, and stiff muscles.

Peduncle: a large bundle of axons which forms the 'stalk' of a neural structure.

Peripheral nervous system: the part of the nervous system outside the central nervous system. This is made up of the somatic and autonomic nerves connecting the central nervous system to other parts of the body.

Pineal gland: a gland attached to the dorsal surface of the thalamus. It releases a chemical called melatonin, which is involved in the sleep-wake cycle.

Piriform cortex: the primary olfactory cortex. The human piriform cortex is located in the uncus.

Pituitary gland: a gland attached to the base of the hypothalamus. It has two parts, the anterior and the posterior, that secrete chemicals involved in the endocrine signaling system.

Plaque: a term used to describe scars in the central nervous system that result from patches of inflammation in multiple sclerosis or protein deposition in Alzheimer's disease.

Plasticity: the ability of the brain to make new connections and to reshape existing connection patterns, thus changing function.

Poliomyelitis: an infectious disease caused by the polio virus, which affects the anterior horn cells of the spinal cord and motor nuclei of the cranial nerves. Symptoms can range from a mild flu-like illness to severe paralysis.

Pons: a collection of nuclei in the ventral part of the hindbrain, from which axons project to the cerebellum.

Posterior: towards the back.

Pretectum: the part of the diencephalon that joins the thalamus to the midbrain. The pretectum is involved in visual reflexes. The landmark for the pretectum is the posterior commissure.

Prethalamus: the part of the diencephalon that lies rostral and ventral to the thalamus and dorsal to the hypothalamus.

Prion: an infectious protein that can cause neurological disease in humans and animals.

Prosencephalon: the embryonic precursor of the forebrain and the cerebral hemispheres.

Prosomere: a segmental component of the developing diencephalon. The three prosomeres are the pretectum (prosomere 1), the thalamus (prosomere 2), and the prethalamus (prosomere 3).

Purkinje cell: the large output cell of the cerebellar cortex. The axons of Purkinje cells project to the cerebellar nuclei.

Pyramidal cell: the predominant cells of the cerebral cortex, named for their pyramid-shaped cell body.

Pyramids: large bundles of corticospinal axons on the ventral surface of the hindbrain next to the midline.

R

Receptor: a specialized part of a cell surface that detects changes in the external environment and initiates changes in cell activity.

Reflex: an involuntary movement in response to a specific stimulus.

Reticular formation: a series of nuclei in the core of the midbrain and hindbrain that have motor and sensory connections.

Rhombencephalon: the embryonic precursor of the hindbrain.

Rhombomere: one of the eleven distinct segments in the embryonic hindbrain, caudal to the isthmus.

Rostral: towards the nose, away from the tail.

S

Saccule: a vesicular region in the labyrinth that contains a patch of hair cells that are sensitive to head position.

Schizophrenia: a psychiatric illness that is characterized by emotional disconnection, bizarre behavior, delusions, and hallucinations.

Schwann cell: glia of the peripheral nervous system that form the myelin sheaths of axons, as well as influencing axonal transport.

Semicircular canals: vestibular sense organs that detect acceleration of the head in any direction. The hair cells in each semicircular canal form a ridge (crista) that is located in an expanded area called the ampulla.

Sensitive period: when the development of a brain system is easily influenced by the external environment.

Sensory nerve: a nerve that transmits signals from sensory receptors to the central nervous system.

Somatic: referring to the skin, muscles and skeleton.

Somatosensory cortex: a granular area of the cortex involved with processing of sensory information. In humans this is found in the post-central gyrus.

Somatotopic organization: the maintenance of topographic relationships between neuronal structures representing different parts of the body.

Spina bifida: a congenital abnormality caused by defective closure of the neural tube.

Spinal cord: the part of the central nervous system located inside the vertebral column.

Spinal nerve: bundles of nerve fibers that extend from the spinal cord to innervate different parts of the body.

Spinal nerve root: each spinal nerve has a ventral and a dorsal root; the ventral root is motor and the dorsal root is sensory.

Spinothalamic tract: bundle of fibers that transmit pain and temperature sensation from the body through the spinal cord to the thalamus.

Stem cell: a precursor cell able to generate many different cell types.

Stereotaxic coordinates: a three-dimensional coordinate system for precise identification of the position of parts of the brain.

Striatum: a deep mass of gray matter in the cerebrum. The main parts of the striatum are the caudate nucleus, the putamen, and the accumbens nucleus. The striatum is involved in motor control.

Stroke: a sudden focal neurological deficit caused by a blocked blood vessel or a hemorrhage into the brain.

Substantia nigra: a cell group in the ventral part of the brainstem that is important in motor control. Many cells in the substantia nigra contain dopamine, and give rise to a dopaminergic pathway which projects to the striatum.

Sulcus: the fissure between folds (gyri) of the cerebral cortex.

Superior colliculus: a raised area of the dorsal surface of the midbrain. Involved in the processing of vision.

Superior olive: a collection of brainstem nuclei involved in auditory processing.

Survival system: brain components whose function is to control survival behaviors, such as feeding, defense, and reproduction. The hypothalamus plays a large role in this system.

Sympathetic nervous system: a division of the autonomic nervous system involved with emergency fight-or-flight responses.

Synapse: a place of close contact between the terminal branch of an axon and another cell. Messages are transmitted across the synapse by neurotransmitters.

T

Tectum: the roof structures of the midbrain, including the superior colliculus and the inferior colliculus.

Thalamus: the major part of the diencephalon, it receives input from touch, auditory, and visual pathways (among others), and sends fibers to the cerebral cortex.

Tract: a group of axons running in a common path in the central nervous system.

Trigeminal nerve: the fifth cranial nerve, which supplies the muscles of the jaw and sensation to the skin of the face.

Trisynaptic circuit: the major neuronal pathway in the hippocampus.

Trochlear nerve: the fourth cranial nerve, which supplies the superior oblique muscle of the eye.

U

Uncus: a gyrus of the temporal lobe, which in humans is the location of the primary olfactory cortex.

Utricle: a vesicular region in the labyrinth that contains a patch of hair cells that are sensitive to head position.

V

Vagus nerve: the tenth cranial nerve, which supplies the muscles of the pharynx and larynx as well as the motor and sensory supply to the internal organs of the thorax and abdomen.

Ventral: towards the belly; away from the back.

Ventral horn: the ventral area of gray matter of the spinal cord that contains motor neuron cell bodies.

Ventricular system: the system of cavities within the brain, which are filled with cerebrospinal fluid.

Vertebra: a bone of the spine, the collection of which makes up the vertebral column.

Vertigo: a false experience of spinning, often accompanied by nausea.

Vestibule: a part of the inner ear (labyrinth) which contains the sense organs of the balance system–the semicircular canals and the saccule and utricle.

Vestibulocochlear nerve: the eighth cranial nerve, which carries the sensations of hearing and balance.

Visceral: a term referring to internal organs, blood vessels, and glands.

Visual system: parts of the nervous system involved in the sense of sight.

W

White matter: brain matter that is principally made up of axons. Myelin is responsible for the whitish color of tracts such as the corpus callosum and pyramidal tract in fresh brain sections.

Bibliography

Allman JM (1998) *Evolving Brains*. Scientific American Library: New York.

Anderson P, Morris R, Amaral D, Bliss T, O'Keefe J (eds.) (2007) *The Hippocampus Book*. Oxford University Press: New York.

Ashwell KWS, Paxinos G (2008) *Atlas of the Developing Rat Nervous System*. 3rd Edition. Academic Press: San Diego.

Barrett L, Henzi P (2005) The social nature of primate cognition. Proc. R. Soc. B 272, 1865-1875.

Bear MF, Connors BW, Paradiso MA (2007) *Neuroscience: Exploring the Brain*. 3rd Edition. Lippincott Williams and Wilkins: New York.

Bowsher D (1967) *Introduction to the Anatomy and Physiology of the Nervous System*. Blackwell Scientific Publications: Oxford.

Brodal A (1981) *Neurological Anatomy in relation to clinical medicine*. 3rd Edition. Oxford University Press: London.

Brodal P (2004) *The Central Nervous System: structure and function*. 3rd Edition. Oxford University Press: New York.

Bull ND, Bartlett PF (2005) The adult mouse hippocampal progenitor is neurogenic but not a stem cell. J Neurosci 25, 10815-10821.

Campbell NA (1996) *Biology*. 4th Edition. Benjamin/Cummings Publishing Company Inc: California.

Carter M, Shieh J (2009) *Guide to Research Techniques in Neuroscience*. Elsevier Academic Press: San Diego.

Citri A, Malenka RC (2008) Synaptic plasticity: multiple forms, functions, and mechanisms. Neuropsychopharm 33, 18-41.

Cramer GD, Darby SA (2005) *Basic and Clinical Anatomy of the Spine, Spinal cord and ANS*. 2nd Edition. Elsevier Mosby: Missouri.

Dan Y, Poo MM (2006) Spike timing-dependent plasticity: from synapse to perception. Physiol Rev 86, 1033-48.

De Felipe J (2010) *Cajal's Butterflies of the Soul: Science and Art*. Oxford University Press: New York.

Desmurget M, Reilly KT, Richard N, Szathmari A, Mottolese C, Sirigu A (2009) Movement intention after parietal cortex stimulation in humans. Science 324, 811-3.

Diekelmann S, Born J (2010) The memory function of sleep. Nat Rev Neurosci 11, 114-26.

Dimond S (1972) *The Double Brain*. The Whitefriars Press Ltd, London.

Disbrow E, Hinckley L, Krubitzer L, Padberrg J (2007) Handuse and the evolution of posterior parietal cortex in primates. In *Evolution of Nervous Systems*. Volume 4–Primates.

Doidge N (2008) *The brain that changes itself*. Scribe Publications: Carlton North.

Donaghy M (2009) *Brain's Diseases of the Nervous System*. 12th Edition. Oxford University Press: New York.

Draganski B, Gaser C, Kempermann G, Kuhn HG, Winkler J, Büchel C, May A (2006) Temporal and spatial dynamics of brain structure changes during extensive learning. J Neurosci 26, 6314-6317.

Dum RP, Strick PL (2002) Motor areas in the frontal lobe of the primate. Physiol Behav 77, 677-82.

Farris EJ, Griffith JQ (1949) *The Rat in Laboratory Investigation*. Hafner Publishing Company: New York.

Franklin KBJ, Paxinos G (2007) *The Mouse Brain in Stereotaxic Coordinates*. 3rd Edition. Elsevier Academic Press: San Diego.

Furness JB (2006) *The Enteric Nervous System*. Wiley-Blackwell: Oxford.

Gazzaniga MS, Sperry RW (1966) Simultaneous double discrimination response following brain bisection. Psychon Sci 4, 261-262.

Glickstein R (2007) What does the cerebellum really do? Current Biol 17, R824-827.

Graziano MSA, Taylor CS, Moore T (2002) Complex movements evoked by microstimulation of precentral cortex. Neuron 34, 841-851.

Haddon M (2003) *The Curious Incident of the Dog in the Night-time*. Jonathon Cape: London.

Herculano-Houzel S (2009) The human brain in numbers: a linearly scaled-up primate brain. Human Neurosc 3, 1-11.

Hofstadter DR, Dennett DC (1981) *The Mind's I: Fantasies and Reflections on Self and Soul*. Basic Books, New York.

Kaas JH (2007) The evolution of sensory and motor systems in primates. In *Evolution of Nervous Systems*. Volume 4–Primates. Kaas JH and Preuss TM (eds) pp35-57. Elsevier/Academic Press: San Diego.

Kandel ER, Schwartz JH, Jessell TM (2000) *Principles of Neural Science*. 4th Edition. McGraw-Hill: New York.

Kettenman H, Ransom B (1995) *Neuroglia*. Oxford University Press: Oxford.

Kirkcaldie MTK, Kitchener PD (2007) When brains expand: mind and the evolution of the cortex. Acta Neuropsych 19, 139-148.

Kötter R, Stephan KE (1997) Useless or helpful? The concept "limbic system". Rev in Neurosci 8, 139-145.

LeDoux J (1998) *The emotional brain: the mysterious underpinnings of emotional life*. Simon and Schuster: New York.

LeDoux J (2007) The amygdala. Current Biology 17, R868-874.

Lemon RN, Johansson RS, Westling G (1995) Corticospinal control during reach, grasp, and precision lift in man. J Neurosc 15, 6145-6156.

Mai JK, Paxinos G (2004) *The Human Nervous System.* 2nd Edition. Elsevier Academic Press: San Diego.

Mai JK, Paxinos G, Voss T (2008) *Atlas of the Human Brain.* 3rd Edition. Elsevier Academic Press: San Diego.

Martin JH (2003) *Neuroanatomy.* 3rd Edition. McGraw-Hill, New York.

MDS Analytical Technologies (2008) *The Axon Guide: A Guide to Electrophysiology & Biophysics Laboratory Techniques* (available online at www.moleculardevices.com/pages/instruments/axon_guide.html).

Nieuwenhuys R, Ten Donkelaar HJ, Nicholson C (1998) *The central nervous system of vertebrates.* Vol 3. Springer, New York.

Oztas E (2003) Neuronal tracing. Neuroanatomy 2, 2-5. *(available online at neuroanatomy.org)*

Paxinos G (2004) *The Rat Nervous System.* 3rd Edition. Elsevier Academic Press: San Diego.

Paxinos G, Halliday G, Watson C, Koutcherov Y, Wang HQ (2007) *Atlas of the Developing Mouse Brain at E17.5, P0, and P6.* Elsevier Academic Press: San Diego.

Paxinos G, Watson C (2007) *The Rat Brain in Stereotaxic Coordinates.* 6th Edition. Elsevier Academic Press: San Diego.

Paxinos G, Watson CRR, Carrive P, Kirkcaldie MTK, Ashwell KWS (2009) *Chemoarchitectonic Atlas of the Rat Brain.* 2nd edition. Elsevier Academic Press: San Diego.

Pourquie O (ed) (2009) *Hox Genes.* Elsevier Academic Press: San Diego.

Rallu M, Corbin JG, Fishell G (2002) Parsing the telencephalon. Nat Rev Neurosci 3, 943-950.

Ramachandran VS, Blakeslee S (1999) *Phantoms in the Brain.* Harper: New York.

Rhodes R (1997) *Deadly feasts.* Simon and Schuster: New York.

Rietze RL, Valcanis H, Brooker GF, Thomas T, Voss AK, Bartlett PF (2001) Purification of a pluripotent neural stem cell from the adult mouse brain. Nature 412, 736-739.

Rizzolatti G, Arbib MA (1998) Language within our grasp. Trends Neurosci 21, 188-194.

Rizzolatti G, Craighero L (2004) The mirror-neuron system. Annu Rev Neurosci 27, 169-92.

Sanes D, Reh TA, Harris WA (2005) *Development of the Nervous System.* 2nd Edition. Elsevier Academic Press: San Diego.

Saper CB, Chou TC, Scammell TE (2001) The sleep switch: hypothalamic control of sleep and wakefulness. Trends Neurosci 24, 726-731.

Saper CB, Scammell TE, Lu J (2005) Hypothalamic regulation of sleep and circadian rhythms. Nature 437: 1257-63.

Scull A (2006) *The Insanity of Place / The Place of Insanity: Essays on the History of Psychiatry.* Routledge: London.

Scull A (2007) *Madhouse: A Tragic Tale of Megalomania and Modern Medicine.* Yale University Press: New Haven.

Sergent C, Baillet S, Dehaene S (2005) Timing of the brain events underlying access to consciousness during the attentional blink. Nature Neurosci 8, 1391-1400.

Shahar A, de Vellis J, Vernadakis A, Haber B (eds.) (1989) *A Dissection and Tissue Culture Manual of the Nervous System.* Alan R Liss Inc: New York.

Shambes GM, Gibson JM, Welker W (1978) Fractured somatotopy in granule cell tactile areas of rat cerebellar hemispheres revealed in micromapping. Brain Behav Evol 15, 94-140.

Siegel JM (2009) Sleep viewed as a state of adaptive inactivity. Nat Rev Neurosci 10, 747-53.

Swanson LW (2003) *Brain Architecture: understanding the basic plan.* Oxford University Press: New York.

Swanson LW (2004) *Brain Maps: Structure of the Rat Brain.* 3rd Edition. Elsevier Inc: San Diego.

Swanson LW, Petrovich GD (1998) What is the amygdala? Trends Neurosci. 21, 323-331.

Tailby C, Wright LL, Metha AB, Calford MB (2005) Activity-dependent maintenance and growth of dendrites in adult cortex. Proc. Natl. Acad. Sci. U.S.A. 102 4631-6.

Turner R, Jones T (2003). Techniques for imaging neuroscience. Br Med Bull 65, 3-20.

Valenstein ES (1973) *Brain Control: A Critical Examination of Brain Stimulation and Psychosurgery.* John Wiley and Sons Inc: New York.

Valenstein ES (1986) *Great and Desperate Cures: The Rise and Decline of Psychosurgery and other Radical Treatments for Mental Illness.* Basic Books: New York.

Watson C, Paxinos G, Kayalioglu G (eds.) (2009) *The Spinal Cord. A Christopher and Dana Reeve Foundation Text and Atlas.* Elsevier Academic Press: San Diego.

Watson CRR, Paxinos G (2010) *Chemoarchitectonic Atlas of the Mouse Brain.* Elsevier Academic Press: San Diego.

Wise SP (2007) Evolution of ventral premotor cortex and the primate way of reaching. In 'Evolution of Nervous Systems'. Volume 4–Primates. Kaas JH and Preuss TM Eds. pp157-165. Elsevier/Academic Press: San Diego.

Woods AE, Ellis RC (1994, 1996) *Laboratory histopathology: a complete reference.* Vols. 1, 2. Churchill Livingstone: Edinburgh.

Woolsey TA, Van der Loos H (1970) The structural organization of layer IV in the somatosensory region (S1) of the mouse cerebral cortex. Brain research 17, 205-242.

Zaborszky L, Wouterlood FG, Lanciego JL (2006) *Neuroanatomical Tract-Tracing: Molecules, Neurons, and Systems.* Springer: New York.

Index

A

abducens nerve	49, 50, 51
accessory nerve	49, 50, 51
accessory olfactory bulb	94
pheromones and	94
accumbens nucleus	37
acetylcholine	5
Alzheimer's disease and	130
autonomic nervous system and	68
acetylcholinesterase	156
action potential	2, 167
adrenal cortex	69
adrenal medulla	69
adrenaline	68
autonomic nervous system and	68
stress and	70, 116
adrenocorticotrophic hormone	72
alar plate	147
Alzheimer's disease	130
amygdala	121
amygdalectomy	137
emotions and	121
inputs	121
memory and	122
outputs	122
social hierarchy and	122
stress and	116, 122
ampulla	91
analgesics	82
anesthetics	82
anterior commissure	36
anxiety	134
aqueduct *see* ventricles	
aqueous chamber	85
arachnoid matter	23
association areas	106
astrocyte	9
development of	9
function of	9
structure of	9
atlas of the brain	26
auditory canal	88
auditory system	87
auditory pathways	89
primary auditory cortex	89
autism	135
autonomic nervous system	51, 68
fight-or flight response	70
functional aspects of	52, 68
neurotransmitters in	52, 53, 70
organization of	52
awake centers	117, 118
sleep control and	117
axon	3
axon myelination *see* myelination	
axon guidance	149

B

Baar's theorem	110
barrel fields	103
basal ganglia	64
basal plate	142
basket cells	99-101
bed nucleus of the stria terminalis *see* stria terminalis	
behavioral modules	59
Bell's palsy *see* facial nerve palsy	
bipolar affective disorder	134
bipolar cell	84
blind spot	84
blood brain barrier	24
Bötzinger complex	120
bovine spongiform encephalopathy (BSE)	127
brain	14
clocks *see* suprachiasmatic nucleus	
death	132
development	142, 152
injury	132
tumor	132
vesicles	14

brainstem..60

branchial (gill) arch................................45

branchial motor neurons....................45, 50

Broca's region.................................105, 107
    language and............................105, 107

Brodmann regions.............................104, 105

BSE *see* bovine spongiform encephalopathy

C

calcium binding proteins.........................158

cauda equina...19

caudate putamen.....................................64

cell culture...160

central canal...23

central nervous system (CNS)....................14

central sulcus..107

cerebellum...15, 61
    anatomy of.......................................61-62
        cortex..62
        connections.................................62
        development of............................62
        folium..62
        functions of................................61
        motor control and.........................61
        vestibular system and....................92

cerebral cortex.......................................98
    behavior and....................................106
    connections....................................102
    development of.................................99
    functional subdivisions............78, 103, 105
    language and...................................107

layers..100
    memory, learning and.........................111
    structure of.....................................100
    vestibular...93

cerebral hemispheres................102-104, 107

cerebral peduncle....................................65

cerebrospinal fluid (CSF).......................10, 23

cerebrovascular accident........................130

cerebrum..98
    hemispheres *see* cerebral hemispheres

chandelier cell......................................100

choroid..84

choroid plexus.......................................23

circadian rhythm.............................118-120

circumventricular organs..........................24

CJD *see* Creutzfeldt-Jakob disease

claustrum...36-37

clutch cells..100

cochlea..89

cochlear nuclei.......................................89

cochlear nerve.......................................88

commissure...19

compulsion..135

computed tomography (CT)....................162

computers...10

cones..84-86

consciousness......................................110
    definition of...................................110
    Baar's theorem and..........................110

cornea...85

corpus callosum................................36-37

corti, organ of.......................................89

corticobulbar tract..................................61

corticopontine tract................................61

corticospinal tract..................................57
    anatomy of......................................57
    injury to..58
    motor control and.............................57

Cotton, Henry......................................138

cranial nerves..47
    anatomy of......................................48
    1st *see* olfactory nerve
    2nd *see* optic nerve
    3rd *see* oculomotor nerve
    4th *see* trochlear nerve
    5th *see* trigeminal nerve
    6th *see* abducens nerve
    7th *see* facial nerve
    8th *see* vestibulocochlear nerve
    9th *see* glossopharyngeal nerve
    10th *see* vagus nerve
    11th *see* accessory nerve
    12th *see* hypoglossal nerve

Creutzfeldt-Jakob disease (CJD)...............127

critical period......................................150

CT *see* computer tomography

CSF *see* cerebrospinal fluid

cuneate fasciculus..................................79

cutaneous receptors................................80

cytochrome oxidase..............................156

D

Delgado, Jose 138
dementia 130
dendrite 3
dentate gyrus 113
    memory and 111
depression 134
    neurotic depression 134
dermatome 47
diffusion tensor imaging (DTI) 162
digestion 54, 71
dopamine 5
    Parkinson's disease and 129
dorsal column medial lemniscus pathway 79
dorsal columns 79
dorsal horn 21
dorsal lateral geniculate nucleus 87
dorsal root ganglion 21
dorsal tegmental nuclei 118
DTI *see* diffusion tensor imaging
dura matter 23

E

ear 88
external 88
    inner 88
    middle 88
ECT *see* electroconvulsive therapy
EEG *see* electroencephalogram
effector system 56
electroconvulsive therapy (ECT) 138
electroencephalography (EEG) 165
electrophysiology 164
emotion 121
    amygdala and 121
    hypothalamus and 122
endocrine system 56
    hypothalamus and 72
    pituitary and 72
endolymph 89
enteric nervous system 56, 71
entorhinal cortex 32-33, 114
ependymal cells 10

epilepsy 132
    temporal lobe epilepsy 132
    seizures 132
    structures and 132
equilibrium potential 166
Eustachian tube 88
evolution 108
    cerebral cortex expansion and 108
external capsule 36
extrapyramidal system 170
eye 84-86
    movements of 49

F

facial nerve 49, 50, 51
    palsy 51
facial nucleus 16
fate mapping 161
fiber stain 155
fight-or-flight 70
    autonomic nervous system and 70
follicle stimulating hormone 72
forebrain *see* prosencephalon
fornix 40
fourth ventricle *see* ventricles
Freud, Sigmund 136
Freudian therapy 136
frontal lobe 103
frontal lobotomy 136
functional imaging 163

G

GABA 5
ganglion 21
ganglion cell (retina) 84
gap junctions 7
gene 142, 161
gene expression 150
    environmental influences and 150
gill arches 45
glia cells 8
    astroglia 9
    functions of 8
    microglia 9
    oligodendroglia 9
    Schwann cell 9

glial markers ... 158
globus pallidus ... 35-36
glossopharyngeal nerve ... 49, 50, 51
glutamate ... 5
Golgi silver impregnation ... 155
Golgi tendon organ ... 68
gracile fasciculus ... 79
granule cells (of cerebellum) ... 62-63
gray matter ... 20, 21
    cerebrum and ... 20
    spinal cord and ... 21
growth cone ... 148
growth hormone ... 72
gyrus ... 104

H

hair cells ... 88-93
headache ... 83
hearing ... 87
herpes zoster ... 47
hindbrain *see* rhombencephalon
hippocampus ... 113
    connections of ... 115
    learning and ... 111
    memory and ... 111
    stress and ... 114
    structure of ... 113
Hirschsprung's disease ... 72
histamine ... 119
histochemical stain ... 155
hodology ... 160
hox genes ... 147
Huntington's disease ... 65
hydrocephalus ... 23
hypoglossal nerve ... 49, 50, 51
hypothalamus ... 59
    endocrine system and ... 72
    function of ... 59
    pituitary gland and ... 72
    survival behaviors and ... 59

I

immunohistochemistry ... 156
iodine deficiency ... 151

incus ... 88
infection ... 126
    of the brain ... 126, 127
    of the spinal cord ... 126
inferior colliculus ... 32
    auditory pathway and ... 89
inferior olive ... 16, 63
    cerebellum and ... 62, 63
inner ear ... 88
in-situ hybridization ... 161
insula ... 105-106
    taste sensation and ... 93
intermediolateral column ... 68
internal capsule ... 65
isthmus ... 40-41

J

joint receptors ... 79-80

K

Kuru ... 127

L

language ... 107
    Broca's area and ... 107
lateral geniculate nucleus *see* dorsal lateral geniculate nucleus
lateral fissure ... 104
lateral ventricles *see* ventricles
L-DOPA ... 5
    Parkinson's disease and ... 129
learning ... 111
    hippocampus and ... 113
    motor learning ... 112
    plasticity and ... 111-112
lectin stains ... 159
lens ... 85
limbic system ... 123
lobotomy ... 136
locomotion ... 60, 65
locus coeruleus ... 31, 61
    sleep control and ... 118
luteinizing hormone ... 72
luxol fast blue stain ... 154

M

macula 84, 92
mad cow disease *see* bovine spongiform encephalopathy (BSE)
magnetic resonance imaging (MRI) 165
magnetoencephalography (MEG) 165
malleus 88
mammillary body 34
mechanoreceptors 79
medial geniculate nucleus 33
    auditory pathway and 89
medial lemniscus 79
median eminence 72
MEG *see* magnetoencephalography
melatonin 32
membrane potential 2, 166
memory 111
    cortex and 112, 113
    episodic 113
    hippocampus and 113
    motor learning and 112
    plasticity and 112
    semantic 116
memory 23
mental illness 133
    historical aspects 136
mesencephalon 14
    anatomy of 14
metal stains 156
microglia 9
microtome 154
middle ear 88
midbrain *see* mesencephalon
migraine 83
migration neurons 144
modules 60
molecular genetics 161
monoculture 156
monosynaptic reflex 62, 95
motor cortex 57-58
    primary motor cortex 57-58
    secondary motor cortex 57-58
supplementary motor cortex 58
motor learning *see* memory 112

motor nerve 44
motor neuron 66
motor systems 56
    basal ganglia and 64
    cerebellum and 61
    motor pathways and 61
motor unit 66
MRI *see* magnetic resonance imaging
multiple sclerosis 127
muscle memory 112
muscle spindle 67
muscle striated 57
muscle stretch receptors 67
myelination 9
myelin stains 155

N

NADPH-D 156
narcolepsy 117
neocortex 99
    development of 144
    evolution of 108
    function of 98
    layers of 101
nerve cell *see* neuron
nerve terminal 4
NeuN 159
neural crest 143
neural groove 142
neural plasticity *see* plasticity
neural tube 142
    formation 142
    neural tube defect 151
neuralgia 83
neuroendocrine system 72
    hypothalamus and 72
    pituitary and 72-74
neurofilaments 158
neuron 2
    death of 149
    development of 144
    migration of 144
    structure of 2
    types of 3
neuroplasticity *see* plasticity
neuroses 134

neurotic depression *see* depression............134

neurotransmitter *see* also specific................5
    actions of..6
    receptors for..5
    synapse and...6

Nissl stain...154

nociception...82
    pain perception and............................82
    pathways..80
    sensitization.......................................83

nodes of Ranvier...9

nor-adrenaline..68

O

obsession..135

obsessive-compulsive disorder (OCD).........135

occipital lobe..103

OCD *see* obsessive compulsive disorder

oculomotor nerve...............................49, 50
    palsy..51

oculomotor nuclei..............................33, 49

olfactory bulb..94

olfactory cortex *see* piriform cortex

olfactory epithelium................................93

olfactory nerve..........................48, 50, 51

olfactory system......................................93
    anatomy of..94
    pathways..94
    pheromones and................................94

olfactory tubercle....................................94

olfactory ventricle...................................94

oligodendrocyte..9

opiates..84
    endogenous production of..................83

opsins..84

optic chiasm..87

optic nerve................................48, 50, 51

optic tract...87
    anatomy of..87
    visual pathway and............................87

organ of Corti..89

organizing centers..................................147
    isthmic organizer..............................147
    sonic hedgehog..........................147, 148

outer ear..88

oxytocin...73

P

pain...80
    nociception and.................................80
    pain receptors....................................82
    pathological pain................................83
    referred pain......................................83
    treatment...84
    visceral pain......................................83

pallidum..64

paracrine...56

parasympathetic nervous system...............53
    anatomy of............................52, 68-70
    function of...70
    neurotransmitters in..........................70
    sympathetic nervous system and.......70

paraventricular nucleus............................72

parietal lobe..103

Parkinson's disease................................129
    dopamine and..................................129
    L-DOPA treatment and.....................129
    stem cell treatment and...................129
    substantia nigra and........................129

periaqueductal gray..................................32

perilymph..88

peripheral nervous system........................44
    cranial nerves....................................47
    motor components............................45
    sensory components..........................45
    somatic components..........................45
    spinal nerves.....................................46
    visceral components.........................45

PET *see* positive emission tomography

Pheromones...94
    accessory olfactory pathway and.......94

photoreceptor..85

pineal gland..32

piriform cortex...94
    anatomy of...................................34-38
    olfactory system and.........................94

pituitary gland...16
    anatomy of...................................73-74
    anterior lobe......................................72
    hormone production and...................72
    hypothalamus and.............................72
    posterior lobe....................................72

plasticity..151
    neuronal..151
    synaptic...151

poliomyelitis..126

polydendocyte 10
pons 32, 64
pontine nuclei 32, 64
portal blood system 73-74
position emission tomography (PET) 164
postcentral gyrus 107
postganglionic neuron 70-71
precentral gyrus 107
preganglionic neuron 70-71
preoptic area 119
pretectal area 34
pretectum *see* pretectal area
prion 127
prolactin 72
proneural genes 144
proprioception 79
prosencephalon 14
prosomeres 34, 145
psychosurgery 136
pupil 85
Purkinje cell 62
pyramidal cell 99-101
pyramidal decussation 28
pyramidal tract 57
pyramids 29

R

rabies 126
radial migration 144
raphe 61
rapid-eye-movement (REM) sleep 117
red nucleus 61
receptors 6
reflex 66-68
    monosynaptic reflex 60, 66
    pain reflex 60
    stretch reflex 67
resting potential 167
reticular activating system 170
reticular formation 28-32
reticular nuclei 28-32
retina 84
    connections 86-87

    signal transduction in 85
    structure of 84
rhombencephalon 28-32
rhombomeres 40, 145, 147
rods 84-86
rostral migratory stream 94
rubella 151
rubrospinal tract 61

S

saccule 90, 92
scala media 89
scala tympani 89
scala vestibuli 89
schizophrenia 133
Schwann cell 9
sclera 85
scrapie 127
semicircular canal 91
sensitive period 150
sensory cortex 78-79
sensory nerve 44
    dorsal root ganglion and 44
sensory systems 76
    comparison of sensory systems 77
    dorsal column medial lemniscus
        pathway and 79
    sensory processing 95
    spinothalamic pathway and 80
septic focus theory 138
septum 36-37
serotonin 5
    mental illness and 134
serotonin-raphe system 61
shingles 47
sleep 117
    control of 117
    learning, memory and 117
    locus coeruleus and 118
    raphe nuclei and 118
    reticular formation and 118
    sleep switch 119
slice culture 160
smell 93
SMI-32 158

social hierarchy 123
    amygdala and............................. 122

solitary nucleus 29

somatic system.................................. 45

somatic nerve 45

somatosensory cortex 81, 105
    connections................................. 81
    primary somatosensory cortex 81

somatosensory system 79
    dorsal column medial lemniscus
        pathway and 79
    proprioception and 79
    pain sensation and......................... 80
    spinothalamic pathway and.................. 80
    temperature sensation and.................. 80
    touch sensation and 79

somatotopic organization 77
    motor cortex and........................... 57
    somatosensory cortex and.................. 81

spasticity 65

spina bifida..................................... 151

spinal cord 19
    anatomy of............................. 19-20
    development of 142
    spinal nerves and 21

spinal nerve 21, 46
    spinal nerve root 46

spinal trigeminal nucleus................. 28-30

spinothalamic pathway 80
    course of 82
    function of................................. 80
    pain sensation and......................... 80
    touch sensation and 80

spinothalamic tract *see* spinothalamic
    pathway

spiny stellate cells......................... 99-101

split-brain...................................... 110
    consciousness and........................ 110

stains 154-155

stapes.. 88

sterotaxic placement 161

strabismus 51

stress pathways 116
    amygdala and......................... 121-122
    endocrine system and 116
    hippocampus and 114
    prolonged stress 116

stria terminalis............................. 36, 40

striatum... 64

stroke 130

subarachnoid space 23
    CSF circulation and........................ 23

subiculum...................................... 114

substance P...................................... 5

substantia nigra................................ 33
    motor control and......................... 66
    Parkinson's disease and 129

sulcus 104

superior colliculus......................... 32-33

superior olive................................... 89

suprachiasmatic nucleus 118
    24-hour clock and........................ 120

supraoptic nucleus 72

survival behaviors............................. 59
    hypothalamus and 59

survival system *see* survival behaviors

sympathetic nervous system 52
    fight-or-flight response and 70
    function of................................. 70
    neurotransmitters in 70
    parasympathetic nervous system and 70

synapse 4
    formation of 148
    neurotransmitters in 4
    plasticity................................. 149
    pruning.................................. 149
    structure of................................. 5

T

tangential migration 144

taste... 93

tectorial membrane............................ 89

temperature..................................... 80
    spinothalamic pathway and................ 80

temporal lobe 103

temporal lobe epilepsy 132

thalamus.................................... 33-35

third ventricle *see* ventricles

three neuron chain............................. 77

thyroid stimulating hormone 72

Timm-Danscher stain 156

TMS *see* transcranial magnetic stimulation

touch 79
    dorsal column medial lemniscus
        pathway and 79
        receptors 76
tracer 160
transcranial magnetic stimulation (TMS) 139
trigeminal nerve 49, 50, 51
trigeminal nucleus 28-30
trisynaptic circuit 114-116
trochlear nerve 49, 50, 51
trochlear nucleus 32
tuberomammillary nucleus 118, 119
tympanic membrane 88

U

uncus 94
    temporal lobe epilepsy and 132
utricle 90-92

V

vagus nerve 49, 50, 51
    anatomy of 49
    functions of 50
    parasympathetic nervous system and 69
vasopressin 74
ventral horn 19-24
ventral tegmental area 33
ventricles 14
    lateral ventricles 14
    third ventricle 14
    fourth ventricle 14
    aqueduct 14

ventrolateral preoptic nucleus 117-119
ventromedial hypothalamic nucleus 35
ventroposterolateral thalamic nucleus 34, 35
ventroposteromedial thalamic nucleus 34
vermis 62
vertebrae 46
vestibular nerve 91
vestibular nuclei 92
vestibulocochlear nerve 49, 50, 51, 90
vestibulospinal tract 92
visceral 45
visceral nerve 45
visual system 84
    primary visual cortex 87
vitreous chamber 85
VLPO *see* ventrolateral preoptic nucleus
vomeronasal organ 94

W

Weigert stain 155
Wernicke's region 107
white matter 20
    of the brain 20
    of the spinal cord 20-22

X

X-ray 162

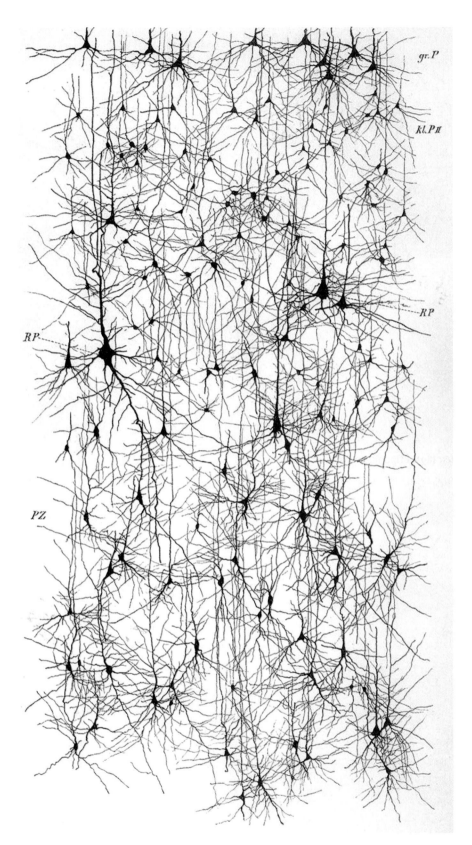

Cortical pyramidal neurons stained with the Golgi method

A diagram of the pyramidal cells of the human cerebral cortex stained with the Golgi method. The original drawing was by Kölliker in 1893. Pyramidal cells have a long apical dendrite and a cluster of basal dendrites. (Image courtesy of De Felipe, 2010, p 88)